Charming Proofs

A Journey into Elegant Mathematics

Library of Congress Catalog Card Number 2010927263
ISBN 978-0-88385-348-1
Printed in the United States of America
Current Printing (last digit):
10 9 8 7 6 5 4 3 2

The Dolciani Mathematical Expositions

NUMBER FORTY-TWO

Charming Proofs

A Journey into Elegant Mathematics

Claudi Alsina
Universitat Politècnica de Catalunya

Roger B. Nelsen
Lewis & Clark College

Published and Distributed by
The Mathematical Association of America

The DOLCIANI MATHEMATICAL EXPOSITIONS series of the Mathematical Association of America was established through a generous gift to the Association from Mary P. Dolciani, Professor of Mathematics at Hunter College of the City University of New York. In making the gift, Professor Dolciani, herself an exceptionally talented and successful expositor of mathematics, had the purpose of furthering the ideal of excellence in mathematical exposition.

The Association, for its part, was delighted to accept the gracious gesture initiating the revolving fund for this series from one who has served the Association with distinction, both as a member of the Committee on Publications and as a member of the Board of Governors. It was with genuine pleasure that the Board chose to name the series in her honor.

The books in the series are selected for their lucid expository style and stimulating mathematical content. Typically, they contain an ample supply of exercises, many with accompanying solutions. They are intended to be sufficiently elementary for the undergraduate and even the mathematically inclined high-school student to understand and enjoy, but also to be interesting and sometimes challenging to the more advanced mathematician.

MAA Service Center
P.O. Box 91112
Washington, DC 20090-1112
1-800-331-1MAA   FAX: 1-301-206-9789

*Dedicated to our many students,
hoping they have enjoyed the beauty of mathematics,
and who have (perhaps unknowingly)
inspired us to write this book.*

Preface

Having perceived the connexions, he seeks the proof, the clean revelation in its
simplest form, never doubting that somewhere writing in the chaos is the unique
elegance, the precise, airy structure, defined, swift-lined, and indestructible.

Lillian Morrison
Poet as Mathematician

Theorems and their proofs lie at the heart of mathematics. In speaking of the "purely aesthetic" qualities of theorems and proofs in *A Mathematician's Apology* [Hardy, 1969], G. H. Hardy wrote that in beautiful proofs "there is a very high degree of *unexpectedness*, combined with *inevitability* and *economy*." These will be the charming proofs appearing in this book.

The aim of this book is to present a collection of remarkable proofs in elementary mathematics (numbers, geometry, inequalities, functions, origami, tilings,...) that are exceptionally elegant, full of ingenuity, and succinct. By means of a surprising argument or a powerful visual representation, we hope the charming proofs in our collection will invite readers to enjoy the beauty of mathematics, to share their discoveries with others, and to become involved in the process of creating new proofs.

The remarkable Hungarian mathematician Paul Erdős (1913–1996) was fond of saying that God has a transfinite Book that contains the best possible proofs of all mathematical theorems, proofs that are elegant and perfect. The highest compliment Erdős could pay to a colleague's work was to say "It's straight from The Book." Erdős also remarked, "You don't have to believe in God, but you should believe in The Book" [Hoffman, 1998]. In 1998 M. Aigner and G. Ziegler gave us a glimpse of what The Book might contain when they published *Proofs from THE BOOK*, now in its second edition [Aigner and Ziegler, 2001]. We hope that *Charming Proofs* complements Aigner and Ziegler's work, presenting proofs that require at most calculus and some elementary discrete mathematics.

But we ask: Are there pictures in The Book? We believe the answer is yes, and you will find over 300 figures and illustrations in *Charming Proofs*. There is a long tradition in mathematics of using pictures to facilitate proofs. This tradition dates back over two thousand years to the mathematics of ancient Greece and China, and continues today with the popular "proofs without words" that appear frequently in the pages of *Mathematics Magazine*, *The College Mathematics Journal*, and other publications. Many of these appear in two books published by the Mathematical Association of America [Nelsen, 1993 and 2000], and we have written two books [Alsina and Nelsen, 2006 and 2009] discussing the process of creating visual proofs, both also published by the MAA.

Charming Proofs is organized as follows. Following a short introduction about proofs and the process of creating proofs, we present, in twelve chapters, a wide and varied selection of proofs we find charming, Each chapter concludes with some challenges for the reader which we hope will draw the reader into the process of creating charming proofs. There are over 130 such challenges.

We begin our journey with a selection of theorems and proofs about the integers and selected real numbers. We then visit topics in geometry, beginning with configurations of points in the plane. We consider polygons, polygons in general as well as special classes such as triangles, equilateral triangles, quadrilaterals, and squares. Next we discuss curves, both in the plane and in space, followed by some adventures into plane tilings, colorful proofs, and some three-dimensional geometry. We conclude with a small collection of theorems, problems, and proofs from various areas of mathematics.

Following the twelve chapters we present our solutions to all of the challenges in the book. We anticipate that many readers will find solutions and proofs that are more elegant, or more charming, than ours! *Charming Proofs* concludes with references and a complete index.

As in our previous books with the MAA, we hope that both secondary school and college and university teachers may wish to use some of the charming proofs in their classrooms to introduce their students to mathematical elegance. Some may wish to use the book as a supplement in an introductory course on proofs, mathematical reasoning, or problem solving.

Special thanks to Rosa Navarro for her superb work in the preparation of preliminary drafts of this manuscript. Thanks too to Underwood Dudley and the members of the editorial board of the Dolciani Mathematical Expositions for their careful reading or an earlier draft of the book and their many helpful suggestions. We would also like to thank Elaine Pedreira, Beverly Ruedi,

Rebecca Elmo, and Don Albers of the MAA's book publication staff for their expertise in preparing this book for publication. Finally, special thanks to students, teachers, and friends in Argentina, New Zealand, Spain, Turkey, and the United States for exploring with us many of these charming proofs, and for their enthusiasm towards our work.

Claudi Alsina
Universitat Politècnica de Catalunya
Barcelona, Spain

Roger B. Nelsen
Lewis & Clark College
Portland, Oregon

Contents

Introduction

The mathematician's patterns, like the painter's or the poet's, must be beautiful; *the ideas, like the colours or the words, must fit together in a harmonious way. Beauty is the first test: there is no place in the world for ugly mathematics.*

G. H. Hardy
A Mathematician's Apology

This is a book about proofs, focusing on attractive proofs we refer to as *charming*. While this is not a definition, we can say that a proof is an argument to convince the reader that a mathematical statement must be true. Beyond mere convincing we hope that many of the proofs in this book will also be fascinating.

Proofs: The heart of mathematics

An elegantly executed proof is a poem in all but the form in which it is written.

Morris Kline
Mathematics in Western Culture

As we claimed in the Preface, proofs lie at the heart of mathematics. But beyond providing the foundation for the growth of mathematics, proofs yield new ways of reasoning, and open new vistas to the understanding of the subject. As Yuri Ivanovich Manin said, "A good proof is one that makes us wiser," a sentiment echoed by Andrew Gleason: "Proofs really aren't there to convince you that something is true—they're there to show you why it is true."

The noun "proof" and the verb "to prove" come from the Latin verb *probare*, meaning "to try, to test, to judge." The noun has many meanings beside the mathematical one, including evidence in law; a trial impression in engraving, photography, printing, and numismatics; and alcohol content in distilled spirits.

Proofs everywhere

At the end of 2009, an Internet search on the word "proof" yielded nearly 24 million web pages, and over 56 million images. Of course, most of these do not refer to mathematical proofs, as proofs are employed in pharmacology, philosophy, religion, law, forensics, publishing, etc. Recently there have been several books, motion pictures, and theatre plays evoking this universal word.

Aesthetic dimensions of proof

> *The best proofs in mathematics are short and crisp like epigrams, and the longest have swings and rhythms that are like music.*
>
> Scott Buchanan
> *Poetry and Mathematics*

What are the characteristics of a proof that lead us to call it charming? In her delightful essay entitled "Beauty and Truth in Mathematics," Doris Schattschneider [Schattschneider, 2006] answers as follows:

- *elegance*—it is spare, cutting right to the essential idea

- *ingenuity*—it has an unexpected idea, a surprising twist

- *insight*—it offers a revelation as to *why* the statement is true, it provides an *Aha!*

- *connections*—it enlightens a larger picture or encompasses many areas

- *paradigm*—it provides a fruitful heuristic with wide application.

Few mathematical terms have attracted as many adjectives as has the word proof. Among the many positive descriptions we find *beautiful, elegant, clever, deep, brief, short, clear, concise, slick, ingenious, brilliant,* and of course *charming*. In the opposite direction, we have *obscure, unintelligible, long, ugly, difficult, complex, lengthy, counterintuitive, impenetrable, incoherent, tedious,* and so on.

One theorem, many proofs

> *Much research for new proofs of theorems already correctly*
> *established is undertaken simply because the existing proofs*
> *have no esthetic appeal. There are mathematical demonstra-*
> *tions that are merely convincing;.... There are other proofs*
> *"which woo and charm the intellect. They evoke delight and*
> *an overpowering desire to say, Amen, Amen."*

> Morris Kline
> *Mathematics in Western Culture*

The importance of a theorem in mathematics often inspires mathematicians
to create a variety of proofs of it. While the statement of the theorem rarely
changes, the existence of a collection of diverse proofs may contribute to a
better understanding of the result, or may open new ways of thinking about
the ideas in question.

The Pythagorean theorem may well be the mathematical theorem with
the greatest number of different proofs. A sequence of twelve articles, each
with the title "New and old proofs of the Pythagorean theorem," appeared
in the *American Mathematical Monthly* between 1896 and 1899, presenting
exactly 100 proofs of the Pythagorean theorem. Building upon this collection and others, Elisha Scott Loomis wrote *The Pythagorean Proposition* in
1907, published in 1927, with a second edition appearing in 1940 that contained 370 proofs. It was re-issued by the National Council of Teachers of
Mathematics in 1968 [Loomis, 1968], and is still a widely cited reference.
New proofs still appear (and old ones reappear) with some regularity.

Once a theorem has been proved, new proofs of the theorem frequently
are published. Murphy's law applied to mathematics implies that the first
proof may be the worst. New proofs may open the possibility of easier arguments, simplified hypotheses, or more powerful conclusions. Proofs of
different types (e.g., algebraic, combinatorial, geometric, etc.) facilitate new
views of the result and fruitful connections between different branches of
mathematics.

In some instances the author of a theorem may give several proofs of
it. For example, Carl Friedrich Gauss (1777–1855) published six different
proofs of the law of quadratic reciprocity (which he called the *aureum the-*
orema, the golden theorem) in his lifetime, and two more were found in
his papers after his death. Today one can find over 200 proofs of this result. In other instances the author of a theorem may fail to publish a proof.
An extreme case is Pierre de Fermat (1601–1665), who wrote in his copy
of *Arithmetica* by Diophantus of Alexandria: "It is impossible to separate a

cube into two cubes, or a fourth power into two fourth powers, or in general, any power higher than the second into two like powers. I have discovered a truly marvelous proof of this, which this margin is too narrow to contain."

Q.E.D. versus the tombstone

Q.E.D., the abbreviation of *quod erat demonstandum* ("which was to be shown") is the Latin translation of the Greek phrase ´οπερ έδει δεîξαι (abbreviated ΟΕΔ), used by Euclid and Archimedes to mark the end of a proof. This form of marking the end of a proof has long been used in English. Its translation into non-Latin languages is also common: in French we have C.Q.F.D. (*ce qu'il fallait démonstrer*), in German w.z.b.w. (*was zu beweisen war*), and in Spanish C.Q.D. (*como queríamos demostrar*).

With the advent of computers and mathematical typesetting software, it has become common to use a geometric shape to mark the end of a proof. Paul Halmos (1916–2006) introduced the tombstone ■ (or ∎), which now competes with the \qed symbol □ in TeX.

The rich world of proof

> *Lightning strikes my mind*
> *I see all, I have the proof*
> *And then I awake.*
>
> Doris Schattschneider
> *A Mathematical Haiku* (after Dante)

Many proofs can be classified according to the methods used in them. Among the most frequent types of proofs are those in the following list. The list is necessarily incomplete, and a proof may often combine several methods.

Direct proof. Use definitions, axioms, identities, inequalities, previously proven lemmas and theorems, etc., to show that the conclusion logically follows from the hypotheses.

Proof by contradiction (also known as *reductio ad absurdum*). Show that it is logically impossible for the statement to be false. This is usually done by assuming the claimed statement to be false and deriving a logical contradiction.

Proof by contrapositive. To prove a conditional statement such as "if A, then B," prove the logically equivalent statement "if not-A, then not-B."

Proof by mathematical induction. A method to show that a statement of the form $P(n)$ is true for every positive integer n: Show that $P(1)$ is true; and show that if $P(n)$ is true, then so is $P(n+1)$.

Proof by cases (or *proof by exhaustion*). Partition the hypothesis into a finite number k of cases, and construct k proofs showing that each case implies the conclusion. The k proofs may be direct, by contradiction, or other methods.

Combinatorial proof. A method of proof for establishing algebraic identities about positive integers (the counting numbers) by representing such numbers by sets of objects and employing one of the following two principles:

1. Counting the objects in a set in two different ways yields the same number, and

2. Two sets in one-to-one correspondence have the same number of elements.

The first principle has been called the *Fubini principle*, after the theorem in multivariable calculus concerning exchanging the order of integration in iterated integrals and the second the *Cantor principle*, after Georg Cantor (1845–1918) who used it extensively in his study of the cardinality of infinite sets. They are also known as the *double-counting method* and the *bijection method*, respectively.

In many cases a proof can be augmented by a figure, and in some instances the figure is sufficient to enable the reader to see the proof. Hence we also consider *visual proof*, or *proof without words*, as a proof technique. For example, a combinatorial proof can be made visual by illustrating the set to be double-counted, or illustrating the bijection between two sets. Other techniques include the use of geometric transformations such as reflection and rotation, changing dimension, tiling, and the use of color. For more on some techniques used to create visual proofs, see our book *Math Made Visual: Creating Images for Understanding Mathematics* [Alsina and Nelsen, 2006].

Proofs in the classroom
Proofs and proving are essential elements of mathematics curricula from elementary schools and high schools to colleges and universities. The National Council of Teachers of Mathematics recommends in its report *Principles*

and Standards for School Mathematics [NCTM, 2000] that instructional programs from pre-kindergarten through 12th grade "should enable all students
to recognize reasoning and proof as fundamental aspects of mathematics."

In its report, the Committee on the Undergraduate Program in Mathematics of the MAA [CUPM, 2004] describes a variety of goals for mathematics
courses for students majoring in the mathematical sciences in colleges and
universities. They state: "The ability to read and write mathematical proofs
is one of the hallmarks of what is often described as mathematical maturity,"
and go on to say that "the foundation for this kind of logical thinking must be
laid in every course in which a prospective mathematics major might enroll,
including calculus and discrete mathematics."

But proofs are not only for students majoring in mathematics. In [CUPM,
2001] the Committee on the Undergraduate Program in Mathematics also
writes: "All students should achieve an understanding of the nature of proof.
Proof is what makes mathematics special." Of course, the CUPM does not
recommend "theorem-proof" type courses for non-mathematics majors, but
they do write that "students should understand and appreciate the core of
mathematical culture: the value and validity of careful reasoning, precise
definition, and close argument."

In a discussion document for the International Commission on Mathematical Instruction [Hanna and de Villiers, 2008], Gila Hanna and Michael de
Villiers write: "A major classroom role for proof is essential to maintaining
the connection between school mathematics and mathematics as a discipline
[since] proof undoubtedly lies at the hear of mathematics." They also note
that "...for mathematicians, proof is much more than a sequence of correct steps, it is also and, perhaps most importantly, a sequence of ideas and
insights, with the goal of mathematical understanding—specifically, understanding why a claim is true. Thus, the challenge for educators is to foster
the use of mathematical proof as a method to certify not only that something
is true, but also why it is true."

It is in this spirit that we have collected and present the proofs in this book.
We hope that you find them not only charming, but also that many readers
will go one step further and find in these pages proofs and challenges to
present in the classroom.

CHAPTER 1

A Garden of Integers

Integers are the fountainhead of all mathematics.
Hermann Minkowski
Diophantische Approximationen

The positive integers are the numbers used for counting, and their use as such dates back to the dawn of civilization. No one knows who first became aware of the abstract concept of, say, "seven," that applies to seven goats, seven trees, seven nights, or any set of seven objects. The counting numbers, along with their negatives and zero, constitute the integers and lie at the heart of mathematics. Thus it is appropriate that we begin with some theorems and proofs about them.

In this chapter we present a variety of results about the integers. Many concern special subsets of the integers, such as squares, triangular numbers, Fibonacci numbers, primes, and perfect numbers. While many of the simpler results can be proven algebraically or by induction, when possible we prefer to present proofs with a visual element. We begin with integers that count objects in sets with a geometric pattern and some identities for them.

1.1 Figurate numbers

The idea of representing a number by points in the plane (or perhaps pebbles on the ground) dates back at least to ancient Greece. When the representation takes the shape of a polygon such as a triangle or a square, the number is often called a *figurate number*. We begin with some theorems and proofs about the simplest figurate numbers: triangular numbers and squares.

Nearly every biography of the great mathematician Carl Friedrich Gauss (1777–1855) relates the following story. When Gauss was about ten years old, his arithmetic teacher asked the students in class to compute the sum $1 + 2 + 3 + \cdots + 100$, anticipating this would keep them busy for some time.

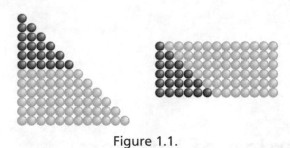

Figure 1.1.

He barely finished stating the problem when young Carl came forward and placed his slate on the teacher's desk, void of calculation, with the correct answer: 5050. When asked to explain, Gauss admitted he recognized the pattern $1 + 100 = 101, 2 + 99 = 101, 3 + 98 = 101$, and so on to $50 + 51 = 101$. Since there are fifty such pairs, the sum must be $50 \cdot 101 = 5050$. The pattern for the sum (adding the largest number to the smallest, the second largest to the second smallest, and so on) is illustrated in Figure 1.1, where the rows of balls represent positive integers.

The number $t_n = 1 + 2 + 3 + \cdots + n$ for a positive integer n is called the nth *triangular number*, from the pattern of the dots on the left in Figure 1.1. Young Carl correctly computed $t_{100} = 5050$. However, this solution works only for n even, so we first prove

Theorem 1.1. *For all $n \geq 1$, $t_n = n(n + 1)/2$.*

Proof. We find a pattern that works for any n by arranging two copies of t_n to form a rectangular array of balls in n rows and $n + 1$ columns. Then we have $2t_n = n(n + 1)$, or $t_n = n(n + 1)/2$. See Figure 1.2.   ∎

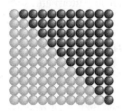

Figure 1.2.

The counting procedure in the preceding combinatorial proof is double counting or the Fubini principle, as mentioned in the Introduction. We employ the same procedure to prove that sums of odd numbers are squares.

Theorem 1.2. *For all $n \geq 1$, $1 + 3 + 5 + \cdots + (2n - 1) = n^2$.*

(a) 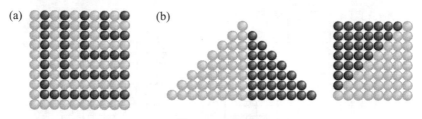 (b)

Figure 1.3.

Proof. We give two combinatorial proofs. In Figure 1.3a, we count the balls in two ways, first as a square array of balls, and then by the number of balls in each L-shaped region of similarly colored balls. In Figure 1.3b, we see a one-to one correspondence (illustrated by the color of the balls) between a triangular array of balls in rows with $1, 3, 5, \ldots, 2n - 1$ balls, and a square array of balls. ∎

The same idea can be employed in three dimensions to establish the following sequence of identities:

$$1 + 2 = 3,$$
$$4 + 5 + 6 = 7 + 8,$$
$$9 + 10 + 11 + 12 = 13 + 14 + 15, \text{ etc.}$$

Each row begins with a square. The general pattern

$$n^2 + (n^2 + 1) + \cdots + (n^2 + n) = (n^2 + n + 1) + \cdots + (n^2 + 2n)$$

can be proved by induction, but the following visual proof is nicer.

In Figure 1.4, we see the $n = 4$ version of the identity where counting the number of small cubes in the pile in two different ways yields $16 + 17 + 18 + 19 + 20 = 21 + 22 + 23 + 24$.

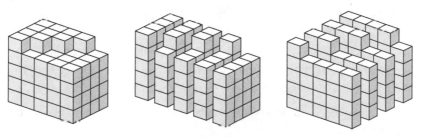

Figure 1.4.

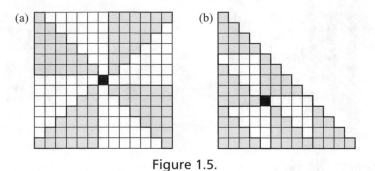

Figure 1.5.

There are many relationships between triangular and square numbers. The simplest is perhaps the one illustrated in the right side of Figure 1.3b: $t_{n-1} + t_n = n^2$. Two more are given in the following lemma (setting $t_0 = 0$ for convenience):

Lemma 1.1. *For all $n \geq 0$, (a) $8t_n + 1 = (2n+1)^2$, and (b) $9t_n + 1 = t_{3n+1}$.*

Proof. See Figure 1.5 (where we have replaced balls by squares). ■

Lemma 1.1 enables us to prove the following two theorems.

Theorem 1.3. *There are infinitely many numbers that are simultaneously square and triangular.*

Proof. From

$$t_{8t_n} = \frac{8t_n(8t_n + 1)}{2} = 4t_n(2n + 1)^2,$$

we see that if t_n is square, then so is t_{8t_n}. Since $t_1 = 1$ is square, this relation generates an infinite sequence of square triangular numbers, e.g., $t_8 = 6^2$, $t_{288} = 204^2$, etc. ■

However, there are square triangular numbers such as $t_{49} = 35^2$ and $t_{1681} = 1189^2$ that are not in this sequence.

Theorem 1.4. *Sums of powers of 9 are triangular numbers, i.e., for all $n \geq 0$, $1 + 9 + 9^2 + \cdots + 9^n = t_{1+3+3^2+\cdots+3^n}$.*

Proof. It is easy to prove this theorem using mathematical induction. The identity in Lemma 1.1b provides the inductive step (see Challenge 1.3). But there is also a nice visual argument, see Figure 1.6. ■

As a consequence, in base 9 the numbers 1, 11, 111, 1111,... are all triangular.

The next theorem presents a companion to the identity $t_{n-1} + t_n = n^2$.

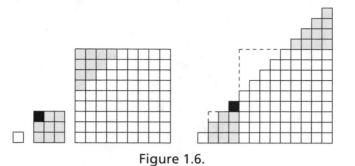

Figure 1.6.

Theorem 1.5. *The sum of the squares of consecutive triangular numbers is a triangular number, i.e., $t_{n-1}^2 + t_n^2 = t_{n^2}$ for all $n \geq 1$.*

Proof. See Figure 1.7, where we illustrate the square of a triangular number as a triangular array of triangular numbers. ∎

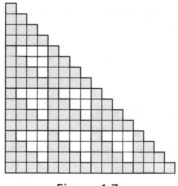

Figure 1.7.

You may have noticed that the nth triangular number is a binomial coefficient, i.e., $t_n = \binom{n+1}{2}$. One explanation for this is that each is equal to $n(n+1)/2$, but this answer sheds little light on why it is true. Here is a better explanation using the Cantor principle:

Theorem 1.6. *There exists a one-to-one correspondence between a set of t_n objects and the set of two-element subsets of a set with $n + 1$ objects.*

Proof. See Figure 1.8 [Larson, 1985], and recall that the binomial coefficient $\binom{k}{2}$ is the number of ways to choose 2 elements from a set of k elements.

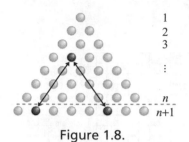

Figure 1.8.

The arrows denote the correspondence between an element of the set with t_n elements and a pair of elements from a set of $n + 1$ elements. ∎

1.2   Sums of squares, triangular numbers, and cubes

Having examined triangular numbers and squares as sums of integers and sums of odd integers, we now consider sums of triangular numbers and sums of squares.

Theorem 1.7. *For all $n \geq 1$, $1^2 + 2^2 + 3^2 + \cdots + n^2 = \dfrac{n(n + 1)(2n + 1)}{6}$.*

Proof. We give two proofs. In the first we exhibit a one-to-one correspondence between three copies of $1^2 + 2^2 + 3^2 + \cdots + n^2$ and a rectangle whose dimensions are $2n + 1$ and $1 + 2 + \cdots + n = n(n + 1)/2$ [Gardner, 1973]. See Figure 1.9.

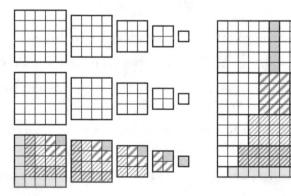

Figure 1.9.

Hence $3(1^2 + 2^2 + 3^2 + \cdots + n^2) = (2n + 1)(1 + 2 + \cdots + n)$ from which the result follows.

In the second proof, we write each square k^2 as a sum of k ks, then place those numbers in a triangular array, create two more arrays by rotating the triangular array by $120°$ and $240°$, and add corresponding entries in each triangular array. See Figure 1.10 [Kung, 1989]. ■

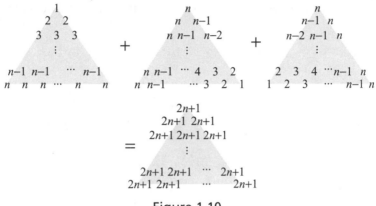

Figure 1.10.

Theorem 1.8. *For all* $n \geq 1$, $t_1 + t_2 + t_3 + \cdots + t_n = \dfrac{n(n + 1)(n + 2)}{6}$.

Proof. In Figure 1.11, we stack layers of unit cubes to represent the triangular numbers. The sum of the triangular numbers is total number of cubes, which is the same as the total volume of the cubes. To compute the volume, we "slice off" small pyramids (shaded gray) and place each small pyramid on the top of the cube from which it came. The result is a large right triangular pyramid minus some smaller right triangular pyramids along one edge of the base.

Thus $t_1 + t_2 + \cdots + t_n = \dfrac{1}{6}(n + 1)^3 - (n + 1)\dfrac{1}{6} = \dfrac{n(n + 1)(n + 2)}{6}$.

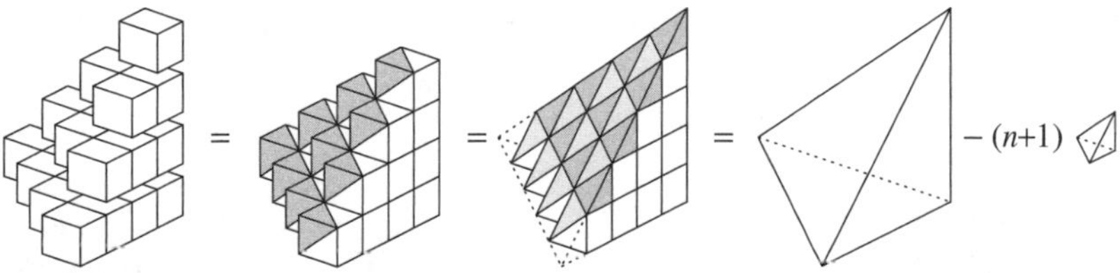

■

Figure 1.11.

In the proof we evaluated the sum of the first n triangular numbers by computing volumes of pyramids. This is actually an extension of the Fubini principle from simple enumeration of objects to additive measures such as length, area and volume. The volume version of the Fubini principle is: *computing the volume of an object in two different ways yields the same number*; and similarly for length and area. We cannot, however, extend the Cantor principle to additive measures—for example, one can construct a one-to-one correspondence between the points on two line segments with different lengths.

Theorem 1.9. *For all $n \geq 1$, $1^3 + 2^3 + 3^3 + \cdots + n^3 = (1 + 2 + 3 + \cdots + n)^2 = t_n^2$.*

Proof. Again, we give two proofs. In the first, we represent k^3 as k copies of a square with area k^2 to establish the identity [Cupillari, 1989; Lushbaugh, 1965].

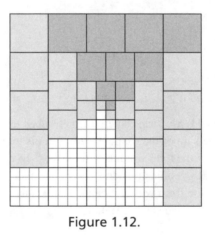

Figure 1.12.

In Figure 1.12, we have $4(1^3 + 2^3 + 3^3 + \cdots + n^3) = [n(n + 1)]^2$ (for $n = 4$).

For the second proof, we use the fact that $1 + 2 + 3 + \cdots + (n - 1) + n + (n - 1) + \cdots + 2 + 1 = n^2$ (see Challenge 1.1a) and consider a square array of numbers in which the element in row i and column j is ij, and sum the numbers in two different ways [Pouryoussefi, 1989].

Summing by columns yields $\sum_{i=1}^{n} i + 2(\sum_{i=1}^{n} i) + \cdots + n(\sum_{i=1}^{n} i) = (\sum_{i=1}^{n} i)^2$, while summing by the L-shaped shaded regions yields (using the result of Challenge 1.1a) $1 \cdot 1^2 + 2 \cdot 2^2 + \cdots + n \cdot n^2 = \sum_{i=1}^{n} i^3$. ∎

1	2	3	$\cdots$	n
2	4	6	$\cdots$	$2n$
3	6	9	$\cdots$	$3n$
$\vdots$	$\vdots$	$\vdots$		$\vdots$
n	$2n$	$3n$	$\cdots$	n^2

1	2	3	$\cdots$	n
2	4	6	$\cdots$	$2n$
3	6	9	$\cdots$	$3n$
$\vdots$	$\vdots$	$\vdots$		$\vdots$
n	$2n$	$3n$	$\cdots$	n^2

Figure 1.13.

We conclude this section with a theorem representing a cube as a double sum of integers.

Theorem 1.10. *For all* $n \geq 1$, $\sum_{i=1}^{n} \sum_{j=1}^{n} (i + j - 1) = n^3$.

Proof. We represent the double sum as a collection of unit cubes and compute the volume of a rectangular box composed of two copies of the collection. See Figure 1.14.

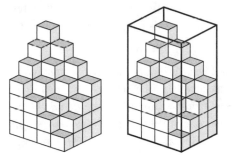

Figure 1.14.

Two copies of the sum $S = \sum_{i=1}^{n} \sum_{j=1}^{n} (i + j - 1)$ fit into a rectangular box with base n^2 and height $2n$, hence computing the volume of the box in two ways yields $2S = 2n^3$, or $S = n^3$. ∎

1.3 There are infinitely many primes

> Reductio ad absurdum, *which Euclid loved so much, is one of a mathematician's finest weapons. It is a far finer gambit than any chess gambit: a chess player may offer the sacrifice of a pawn or even a piece; but a mathematician offers the game.*
>
> G. H. Hardy
> *A Mathematician's Apology*

The earliest proof that there are infinitely many primes is probably Euclid's in the *Elements* (Book IX, Proposition 20). After more than 2000 years, it

is difficult to find a better one. We present three distinctly different proofs.
Proof 1 below, based on one created by Ernst Eduard Kummer (1810–1893)
in 1873, is quite elegant. It is a *reductio ad absurdum* proof. Proof 2 is a
direct proof, and even simpler. Henri Brocard (1845–1922) published it in
1915, attributing it to Charles Hermite (1822–1901) [Ribenboim, 2004]. In
Proof 3 we construct integers with an arbitrary number of distinct prime
factors [Saidak, 2006].

Euclid's Theorem 1.11. *There are infinitely many prime numbers.*

Proof 1. Suppose that there are only k primes, $p_1, p_2, \ldots, p_k$. Let $N = p_1 p_2 \cdots p_k$. Since $N + 1$ is larger than p_k it is not prime and thus has a
prime divisor p_j in common with N. Since p_j divides both N and $N + 1$,
it divides $(N + 1) - N = 1$, which is absurd. ∎

Proof 2. It suffices to show that for every positive integer n, there exists a
prime p greater than n. For this purpose one considers any prime p dividing
$n! + 1$. ∎

Proof 3. Let $n > 1$ be an arbitrary integer. Since n and $n + 1$ are consecutive
integers, they are relatively prime. Thus $N_2 = n(n + 1)$ must have at least
two different prime factors. Similarly since $n(n + 1)$ and $n(n + 1) + 1$ are
consecutive and hence relatively prime, $N_3 = n(n + 1)[n(n + 1) + 1]$ must
have at least three different prime factors. This process can be continued
indefinitely. ∎

Euclid primes

Numbers of the form $N_k = p_1 p_2 \cdots p_k$ are called *primorials* (from
prime and *factorial*), and the number $E_k = N_k + 1$ is called a *Euclid
number*. The first five Euclid numbers $3, 7, 31, 211, 2311$ are prime (and
called *Euclid primes*), however $E_6 = 30031 = 59 \cdot 509$. It is not known
if the number of Euclid primes is finite or infinite.

In 1737 Leonhard Euler (1707–1783) proved that there are infinitely many
primes by showing that a certain expression involving all the primes is infinite. One such expression is the sum of the reciprocals of the primes; if it is
infinite, then there must be infinitely many primes. Here is a modern proof
of that result by F. Gilfeather and G. Meisters [Leavitt, 1979] using only
calculus and the divergence of the harmonic series, which we now prove.

Lemma 1.2. *The harmonic series* $1 + \dfrac{1}{2} + \dfrac{1}{3} + \cdots$ *diverges.*

Proof [Ward, 1970]. Let $H_n = 1 + (1/2) + (1/3) + \cdots + (1/n)$ denote the nth partial sum of the harmonic series. Assume the harmonic series converges to H. Then $\lim_{n\to\infty}(H_{2n} - H_n) = H - H = 0$. But

$$H_{2n} - H_n = \frac{1}{n+1} + \frac{1}{n+2} + \cdots + \frac{1}{2n} > n \cdot \frac{1}{2n} = \frac{1}{2},$$

so that $\lim_{n\to\infty}(H_{2n} - H_n) \neq 0$, a contradiction. ∎

Theorem 1.12. $\sum_{p \text{ prime}} 1/p$ *diverges.*

Proof. For a fixed integer $n \geq 2$, consider the set of all primes $p \leq n$ and the product

$$\prod_{p \leq n} \left(\frac{p}{p-1} \right) = \prod_{p \leq n} \left(\frac{1}{1 - 1/p} \right) = \prod_{p \leq n} \left(1 + \frac{1}{p} + \frac{1}{p^2} + \cdots \right).$$

Since each number $k \leq n$ is a product of powers of primes $p \leq n$, it follows that for each $k \leq n$, $1/k$ must appear as one of the terms in the product on the right. Hence

$$\prod_{p \leq n} \left(\frac{p}{p-1} \right) > \sum_{k=1}^{n} \frac{1}{k}.$$

Since the natural logarithm is an increasing function it preserves inequalities, hence

$$\sum_{p \leq n} [\ln p - \ln(p-1)] > \ln \left(\sum_{k=1}^{n} \frac{1}{k} \right). \tag{1.1}$$

However,

$$\sum_{p \leq n} [\ln p - \ln(p-1)] = \sum_{p \leq n} \left(\int_{p-1}^{p} \frac{1}{x} dx \right) < \sum_{p \leq n} \left(\frac{1}{p-1} \right) \leq \sum_{p \leq n} \frac{2}{p}. \tag{1.2}$$

Combining (1.1) and (1.2) yields

$$\sum_{p \leq n} \frac{1}{p} > \frac{1}{2} \ln \left(\sum_{k=1}^{n} \frac{1}{k} \right). \tag{1.3}$$

The right side of (1.3) increases without bound as $n \to \infty$, and hence $\sum_{p \text{ prime}} 1/p$ diverges. ∎

For additional proofs, see [Vanden Eynden, 1980].

Primes and security

A rewarding aspect of mathematics is the sometimes unexpected application of some branch of mathematics. For centuries the study of prime numbers was motivated by basic arithmetic questions. But now, in our digital society where all computer communication is based on the exchange of numbers, the prime numbers have become an essential tool for security. In today's cryptography many methods (RSA of R. Rivest, A. Shamur, and L. Adleman; Elgamal's method; methods of R. Merkle, W. Diffie, M. Hellman, ...) depend on primes. The key idea is that if two large primes p and q are multiplied, recovering p or q from pq can be a very difficult task.

1.4   Fibonacci numbers

Suppose we have a collection of identical unit squares and identical 1-by-2 rectangles we call *dominos* and an *n-board*, a 1-by-n rectangle. Let f_n denote the number of distinguishable ways in which we can *tile* an n-board, that is, place squares and rectangles on the board to cover it with no overlapping of squares and rectangles. For example, $f_5 = 8$—see Figure 1.15 for a picture of a 5-board and the eight ways to tile it with squares and rectangles.

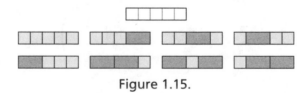

Figure 1.15.

Similarly, we have $f_1 = 1$, $f_2 = 2$, $f_3 = 3$, $f_4 = 5$, $f_5 = 8$, $f_6 = 13$, and so on. In fact, it is easy to see that for any $n \geq 3$, $f_n = f_{n-1} + f_{n-2}$: to tile an n-board, start at the left with a square and complete the tiling in one of f_{n-1} ways, or start at the left with a domino and complete the tiling in f_{n-2} ways.

You have probably noticed that the range of the sequence $\{f_n\}_{n=1}^{\infty}$ is the same as the range of the *Fibonacci sequence* $\{F_n\}_{n=1}^{\infty}$ given by $F_n = F_{n-1} + F_{n-2}$ with $F_1 = F_2 = 1$. Indeed, if we define $f_0 = 1$ (there is just one way to tile the 0-board—no squares, no dominos), then for all $n \geq 0$, $f_n = F_{n+1}$.

To prove identities for the Fibonacci numbers, we need only prove identities for the sequence $\{f_n\}_{n=0}^{\infty}$. The next two theorems and their proofs are from *Proofs That Really Count* [Benjamin and Quinn, 2003], a delightful collection of lovely proofs about the Fibonacci and related sequences. The proofs employ the Fubini principle, counting in two different ways the number of tilings of n-boards with squares and dominos subject to a particular condition, a method Benjamin and Quinn call *conditioning*.

Theorem 1.13. *For $n \geq 0$, $f_0 + f_1 + f_2 + \cdots + f_n = f_{n+2} - 1$.*

Proof. How many tilings of an $(n + 2)$-board use at least one domino? Almost by definition, one answer is $f_{n+2} - 1$, since we exclude the one tiling by all squares from the unrestricted total f_{n+2}. For the second way to count, we look at the location of the leftmost domino (the *condition* in this example for the conditioning method). If the tiling of an $(n + 2)$-board begins on the left with a domino, it can be completed in f_n ways. If the tiling begins on the left with a square and then a domino, it can be completed in f_{n-1} ways. If the tiling begins on the left with two squares and then a domino, it can be completed in f_{n-2} ways. See Figure 1.16.

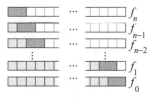

Figure 1.16.

We continue to the final case, wherein the tiling begins with n squares and a domino, which can be completed in $f_0 = 1$ way (that is, it already is a tiling of the $(n + 2)$-board). Summing yields the desired result. ∎

Theorem 1.14. *For $n \geq 0$, $f_0 + f_2 + f_4 + \cdots + f_{2n} = f_{2n+1}$.*

Proof. How many tilings of a $(2n+1)$-board are there? By definition, f_{2n+1}. For a second way to count, we condition on the location of the leftmost square. If the tiling of a $(2n + 1)$-board begins on the left with a square, it can be completed in f_{2n} ways. If the tiling begins on the left with a domino and then a square, it can be completed in f_{2n-2} ways. If the tiling begins on the left with two dominos and then a square, it can be completed in f_{2n-4} ways. See Figure 1.17.

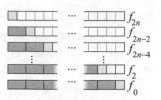

Figure 1.17.

We again continue to the final case, wherein the tiling begins with n dominos and a square, which can be completed in $f_0 = 1$ way. Summing yields the desired result. ∎

See [Benjamin and Quinn, 2003] for many more such Fibonacci identities and combinatorial proofs.

Identities involving powers of Fibonacci numbers such as squares and cubes can be illustrated nicely with two- or three-dimensional pictures.

Theorem 1.15. *For* $n \geq 1$, $F_1^2 + F_2^2 + \cdots + F_n^2 = F_n F_{n+1}$.

Proof. See Figure 1.18 [Bicknell and Hoggatt, 1972]. ∎

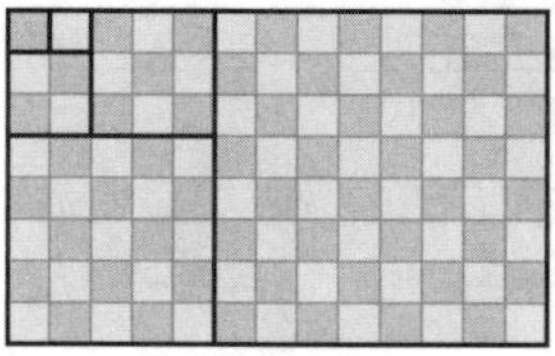

Figure 1.18.

Theorem 1.16. *For* $n \geq 2$, $F_{n+1}^3 = F_n^3 + F_{n-1}^3 + 3F_{n-1}F_nF_{n+1}$.

Proof. See Figure 1.19. ∎

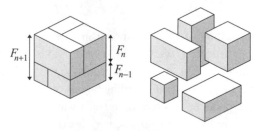

Figure 1.19.

It may be surprising that there is no explicit formula for the Fibonacci numbers using only integers, but there is one using the golden ratio. See Section 2.1 in the next chapter.

Fibonacci numbers everywhere

Leonardo of Pisa (circa 1170–1240) may not have known that he would be called Fibonacci (a contraction of *filius Bonaccio*, son of Bonaccio), and he certainly never dreamt that his sequence 1, 1, 2, 3, 5, ..., which he introduced in a problem about counting rabbits, would become such a celebrated sequence of integers. The journal *The Fibonacci Quarterly*, first published in 1963, is devoted to the study of the properties of this sequence. Fibonacci numbers appear frequently in nature (phyllotaxis, sunflowers, pine cones, pineapples, artichokes, family trees for honeybees, etc.) as well as in architecture and design.

1.5 Fermat's theorem

One of the most useful tools in number theory is the result known as *Fermat's theorem*. It is sometimes called Fermat's "little" theorem to distinguish it from his more famous "great" or "last" theorem. Pierre de Fermat (1601–1665) mentioned it (without proof) in a letter in 1640; Leonhard Euler published the first proof in 1736.

There are many known proofs of Fermat's theorem. The one we present is particularly elegant. In it we simply count objects in a set [Golomb, 1956].

Fermat's Theorem 1.17. *If n is an integer and p a prime, then p divides $n^p - n$. Furthermore, if n is not a multiple of p, then p divides $n^{p-1} - 1$.*

Proof. It suffices to prove the theorem for positive integers (see Challenge 1.7). Suppose we have a supply of beads in n distinct colors, and we wish to make multicolored necklaces consisting of p beads. We first place p beads on a string. Since each bead can be chosen in n ways, there are n^p possible strings of beads. For each of the n colors, there is one string whose beads are all the same color, which we discard, leaving $n^p - n$ strings. We now join the ends of the strings together seamlessly to form necklaces. But we notice that if two necklaces differ only by a cyclic permutation of the beads, then they are indistinguishable. For p prime there are p cyclic permutations of the p beads, so that the number of *distinguishable multicolored* neck-laces is $(n^p - n)/p$, which is therefore an integer. Finally since $n^p - n = n(n^{p-1} - 1)$, if p does not divide n, then p must divide $n^{p-1} - 1$.   ∎

1.6   Wilson's theorem

In terms of congruences, Fermat's theorem is stated as follows: If n is an integer and p a prime, then $n^p \equiv n \pmod{p}$. Furthermore, if n is not a multiple of p, then $n^{p-1} \equiv 1 \pmod{p}$. One of the beautiful consequences of Fermat's theorem is

Wilson's Theorem 1.18. *If p is prime, then $(p-1)! \equiv -1 \pmod{p}$.*

Proof. The theorem is clearly true for $p = 2$, so assume p odd. From Fermat's theorem, the integers $1, 2, \ldots, p-1$ are roots of the congruence $x^{p-1} - 1 \equiv 0 \pmod{p}$. Since a polynomial congruence of degree $p-1$ has exactly $p-1$ incongruent solutions modulo p, we have

$$x^{p-1} - 1 \equiv (x-1)(x-2)\cdots(x-(p-1)) \bmod p.$$

If we compare the constant terms modulo p, we have $-1 \equiv (-1)^{p-1}$ $(p-1)! \equiv (p-1)! \pmod{p}$, as required.   ∎

The converse of Wilson's theorem is also true—if $(n-1)! \equiv -1 \pmod{n}$, then n is prime. To prove this, assume n is not prime, so that $n = ab$ with $1 < a, b < n-1$. Then a divides both n and $(n-1)!$, so $(n-1)! \not\equiv -1 \pmod{n}$.

1.7   Perfect numbers

> *Perfect numbers have engaged the attention of arithmeti-cians of every century of the Christian era.*
>
> L. E. Dickson
> *History of the Theory of Numbers*

> *Perfect numbers like perfect men are very rare.*
>
> René Descartes

A *perfect number* is a positive integer n that is equal to the sum of its positive divisors excluding itself. For example, 6 and 28 are perfect, since $6 = 1 + 2 + 3$ and $28 = 1 + 2 + 4 + 7 + 14$. In Book IX, Proposition 36 of the *Elements*, Euclid tells us how to construct more perfect numbers.

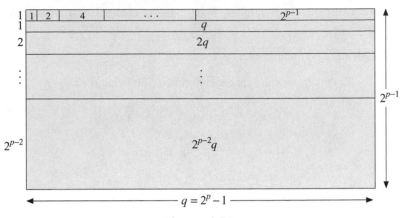

Figure 1.20.

Theorem 1.19. (Euclid). *If p and $q = 2^p - 1$ are prime, then $2^{p-1}q$ is perfect.*

Proof. The divisors of $2^{p-1}q$ are $\{1, 2, 2^2, \ldots, 2^{p-1}, q, 2q, 2^2q, \ldots, 2^{p-2}q\}$. In Figure 1.20 we see how to arrange squares and rectangles whose areas are the divisors into a rectangle with area $2^{p-1}q$ [Goldberg]. ∎

About 2000 years after Euclid, Leonard Euler (1707–1783) proved the converse of Theorem 1.19, that is, every even perfect number must be of the form $2^{p-1}(2^p - 1)$ where p and $2^p - 1$ are prime. It is worth noting that every even perfect number is a triangular number. It is not known if odd perfect numbers exist.

1.8 Challenges

1.1 Use a combinatorial argument to show that

    (a) $1 + 2 + 3 + \cdots + (n-1) + n + (n-1) + \cdots + 2 + 1 = n^2$

    (b) $1 + 3 + 5 + \cdots + (2n-1) + (2n+1) + (2n-1) + \cdots + 3 + 1 = n^2 + (n+1)^2$

    (c) $\sum_{k=1}^{n} k^2 = \sum_{i=1}^{n} \sum_{j=1}^{n} \min(i, j)$.

1.2 Create visual proofs that (a) $3t_n + t_{n-1} = t_{2n}$, (b) $3t_n + t_{n+1} = t_{2n+1}$, and (c) $t_{n-1} + 6t_n + t_{n+1} = (2n+1)^2$.

1.3 Use mathematical induction to prove the identity in Theorem 1.4.

1.4 Prove that there are infinitely many pairs of triangular numbers whose sum is triangular.

1.5 Can a Euclid number ever be a square?

1.6 Let F_n denote the nth Fibonacci number. Show that for $n \geq 2$,

$$
\begin{aligned}
F_{n+1}^2 &= 2F_{n+1}F_n - F_n^2 + F_{n-1}^2 \\
&= 2F_{n+1}F_{n-1} + F_n^2 - F_{n-1}^2 \\
&= 2F_n F_{n-1} + F_n^2 + F_{n-1}^2 \\
&= F_{n+1}F_n + F_n F_{n-1} + F_{n-1}^2 \\
&= F_{n+1}F_{n-1} + F_n^2 + F_n F_{n-1}.
\end{aligned}
$$

[Hint: A single figure suffices for this Challenge.]

1.7 Let F_n denote the nth Fibonacci number. Using illustrations similar to Figure 1.18, show that

$$
\begin{aligned}
F_1 F_3 + F_2 F_4 + \cdots + F_{2n}F_{2n+2} &= F_2^2 + F_3^2 + \cdots + F_{2n+1}^2 \\
&= F_{2n+1}F_{2n+2} - 1.
\end{aligned}
$$

1.8 Let F_n denote the nth Fibonacci number. Prove *Cassini's identity*: for all $n \geq 2$, $F_{n-1}F_{n+1} - F_n^2 = (-1)^n$. [Hint: First show that for all $n \geq 2$, $F_{n+1}^2 - F_n F_{n+2} = F_{n-1}F_{n+1} - F_n^2$. This can be done with a visual argument.]

1.9 A sequence $\{a_n\}$ of positive integers is defined by $a_1 = 1, a_2 = 2$, and the recurrence $a_{n+1} = a_n + 2a_{n-1} + 1$ for $n \geq 3$. Find an explicit expression for a_n. [Hint: Base 10 may not always be the best way to express numbers.]

1.10 Show that if Fermat's theorem 1.17 holds when n is a positive integer, then it holds for all integers n.

1.11 For a positive integer n let $\tau(n)$ denote the number of divisors of n. By definition, n is a prime number if and only if $\tau(n) = 2$. Prove that n is a square number if and only if $\tau(n)$ is odd.

1.12 Let n be an even perfect number greater than 6. Show that (a) n is one more than a multiple of 9 and (b) when $n > 28$, $(n-1)/9$ is never a prime.

CHAPTER 2

Distinguished Numbers

> *Number, the most excellent of all inventions.*
> Aeschylus, *Prometheus Bound*
>
> *Numbers are a fearful thing.*
> Euripides, *Hecuba*
>
> *Wherever there is number, there is beauty.*
> Proclus

Numbers are not only beautiful; some are popular, so popular that they have had their biographies written. Here is a short list of number biographies published since 1994:

e: The Story of a Number [Maor, 1994];

The Joy of π [Blatner, 1997];

The Golden Ratio and Fibonacci Numbers [Dunlap, 1997];

The Golden Ratio: The Story of Phi, the World's Most Astonishing Number [Livio, 2002];

An Imaginary Tale: The Story of i [Nahin, 1998];

Gamma: Exploring Euler's Constant [Havil, 2003]; and

The Square Root of Two [Flannery, 2006].

In this chapter we prove some basic results about some special numbers such as $\sqrt{2}$, π, and e. The proofs we have chosen to present are, like the numbers themselves, considered by many to be beautiful.

Distinguished names for distinguished numbers

Distinguished numbers have not only values, many have names. $\sqrt{2}$ is sometimes called *Pythagoras's constant*, as it was first shown to be irrational by the Pythagoreans. π is called *Archimedes' constant*, as Archimedes first established the inequality $3^{10}/_{71} < \pi < 3^1/_7$. π is sometimes called *Ludolph's constant* for Ludolph van Ceulen (1540–1610), who spent many years computing π to 35 places. *e* is sometimes called *Euler's constant*, but the same name is sometimes used for the *Euler-Mascheroni constant* γ. Since John Napier (1550–1617) almost discovered *e*, it is sometimes called *Napier's constant*.

2.1 The irrationality of $\sqrt{2}$

There are many proofs that $\sqrt{2}$ is irrational. Perhaps the best known is Euclid's proof, based on the Pythagorean theorem. It is perhaps the oldest as well. But there are others that can be considered simpler. Here is a one-sentence version of another old proof of the irrationality of $\sqrt{2}$ [Bloom, 1995]:

Theorem 2.1. $\sqrt{2}$ *is irrational.*

Proof. If $\sqrt{2}$ were rational, say $\sqrt{2} = m/n$ in lowest terms, then $\sqrt{2} = (2n - m)/(m - n)$ in lower terms, giving a contradiction. ∎

To believe this, the reader must verify three facts—(i) the two fractions are equal, (ii) the second denominator is positive, and (iii) the second denominator is smaller than the first—all of which are easily done. While the verification can be done algebraically, it is perhaps more easily understood when seen geometrically [Apostol, 2000]. If $\sqrt{2} = m/n$ in lowest terms, then an isosceles triangle with sides n, n, and m is the *smallest* isosceles right triangle with integer sides. However,

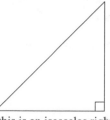

if this is an isosceles right   then there is a smaller one
triangle with integer sides,   with the same property.

Figure 2.1.

Indeed, if the sides of the large triangle above are n, n, and m, then the sides of the shaded triangle on the right are $m - n$, $m - n$, and $2n - m$. Thus the assumption that $\sqrt{2} = m/n$ in lowest terms is false, and $\sqrt{2}$ is irrational.

Hippasus and the irrationality of $\sqrt{2}$

Proving theorems in mathematics is not usually dangerous. Indeed it is often quite rewarding, but in Pythagoras' time, the situation may have been different. Legend has it that Hippasus of Metapontum (c. 500 BCE) discovered the unexpected irrationality of $\sqrt{2}$, explained his result, and consequently was thrown overboard at sea by his fellow Pythagoreans. The discovery had destroyed the ideal of commensurability in geometry. Fortunately, history records no other such deaths as a result of proving theorems.

2.2 The irrationality of $\sqrt{k}$ for non-square k

We can modify the proof of the irrationality of $\sqrt{2}$ to show that $\sqrt{k}$ is irrational when k is not the square of an integer. In this proof, we interpret $\sqrt{k}$ as the slope of a line through the origin, as illustrated in Figure 2.2.

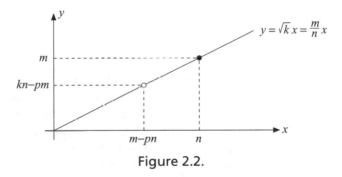

Figure 2.2.

Theorem 2.2. *If k is not the square of an integer, then $\sqrt{k}$ is irrational.*

Proof. Assume $\sqrt{k} = m/n$ in lowest terms. Then the point on the line $y = \sqrt{k}x = (m/n)x$ closest to the origin with integer coordinates is (n,m). However, if we let p be the greatest integer less than $\sqrt{k}$ so that $p < \sqrt{k} < p+1$, then the point with integer coordinates $(m - pn, kn - pm)$ lies on the line and is closer to the origin since $(m/n)(m - pn) = m^2/n - pm = kn - pm$, and $p < m/n < p + 1$ implies $0 < m - pn < n$ and $0 < kn - pm < m$. Thus we have a contradiction and $\sqrt{k}$ is irrational. ■

Since we write numbers in base 10, it is easy to show that $\sqrt{10}$ is irrational. If $\sqrt{10}$ is rational, then $\sqrt{10} = m/n$ in lowest terms. Then we have $m^2 = 10n^2$. But the base 10 representation of a square must end in an even number of zeros, so m^2 must end with both an even and an odd number of zeros, which is impossible.

2.3   The golden ratio

What is the most aesthetically pleasing shape for a rectangle? Some people (but certainly not everyone) would say that such a rectangle has the shape illustrated in Figure 2.3a, where if the shorter side has length 1, the longer side has length $\varphi > 1$. This "golden rectangle" has the property that if we cut off a square from the rectangle (as in Figure 2.3b), the new rectangle is similar to the original, and the process can be continued indefinitely (see Figure 2.3c). Hence, as in Figure 2.3b, φ must satisfy $\varphi/1 = 1/(\varphi - 1)$.

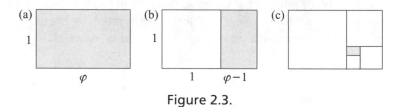

Figure 2.3.

Thus $\varphi^2 = \varphi + 1$. This quadratic equation has two roots, but only one is positive so $\varphi = (1 + \sqrt{5})/2 \approx 1.618$. This root is called the *golden ratio* or the *divine proportion*. Euclid referred to the procedure for constructing line segments in this ratio as "cutting a line into extreme and mean ratio" in Definition 3 and Proposition 30 in Book VI of the *Elements*.

The wonders of the golden ratio

As Mario Livio wrote [Livio, 2002, p. 6], "In fact, it is probably fair to say that the Golden Ratio has inspired thinkers of all disciplines like no other number in the history of mathematics." Since the ancient Greeks discovered the golden ratio in pentagons and in the division of line segments into extreme and mean ratio, there has been an interest in art and architecture for using this ratio for it aesthetic appeal. The three-volume work *Divina Proportione,* published by Luca Pacioli in 1509 and lavishly illustrated by Leonardo da Vinci, promoted the use of the golden ratio in art. In the twentieth century the book *Le Modulor* by the architect Le Corbusier presented a system of proportions based on the golden ratio that continues to influence artists and architects to this day. Finally, the tiles of Roger Penrose, based on isosceles triangles whose sides are in the golden ratio, have led to new mathematical results based on this old number.

Analogous to $\sqrt{2}$ as the length of the diagonal of the unit square, φ is the length of the diagonal of the regular pentagon with side 1, as illustrated in Figure 2.4 (see Challenge 2.4).

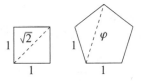

Figure 2.4.

Since $\varphi^2 = \varphi + 1$, it is possible to express every positive integer power of φ in the form $a + b\varphi$ where a and b are integers. We begin by multiplying both sides of $\varphi^2 = \varphi + 1$ by φ and simplifying, and repeat the process:

$$\varphi^3 = \varphi^2 + \varphi = 2\varphi + 1,$$
$$\varphi^4 = 2\varphi^2 + \varphi = 3\varphi + 2,$$
$$\varphi^5 = 3\varphi^2 + 2\varphi = 5\varphi + 3,$$
$$\varphi^6 = 5\varphi^2 + 3\varphi = 8\varphi + 5,$$

and so on. To establish the pattern for the integers a and b, set $\varphi^n = a_{n-1} + b_n\varphi$ for $n \geq 2$ with $a_1 = b_2 = 1$. Then

$$a_n + b_{n+1}\varphi = \varphi^{n+1} = a_{n-1}\varphi + b_n\varphi^2 = b_n + (a_{n-1} + b_n)\varphi,$$

so that $a_n = b_n$ and $b_{n+1} = b_n + b_{n-1}$. Thus $\{b_n\}$ is the sequence $\{F_n\}$ of Fibonacci numbers, since it satisfies the same recurrence and has the same initial values (see Section 1.6). Replacing $\{b_n\}$ by $\{F_n\}$ we have $\varphi^n = F_{n-1} + F_n\varphi$ for $n \geq 2$.

Since $1/\varphi = \varphi - 1$ we can establish a similar relationship for negative integer powers of φ: $(-1)^n \varphi^{-n} = F_{n+1} - F_n \varphi$. Hence

$$
\begin{aligned}
\varphi^n - (-1)^n \varphi^{-n} &= 2F_n\varphi - (F_{n+1} - F_{n-1}), \\
&= 2F_n\varphi - F_n = (2\varphi - 1)F_n, \\
&= \sqrt{5}F_n.
\end{aligned}
$$

Thus we have *Binet's formula*, expressing the Fibonacci numbers in terms of the golden ratio.

Theorem 2.3 (Binet's formula). $F_n = [\varphi^n - (-1)^n \varphi^{-n}]/\sqrt{5}$.

Since $(-1)^n \varphi^{-n} = F_{n+1} - F_n \varphi$ it follows that the ratio of successive Fibonacci numbers has a limit as $n \to \infty$:

$$
\lim_{n \to \infty} \frac{F_{n+1}}{F_n} = \lim_{n \to \infty} \left(\varphi + \frac{(-1)^n}{\varphi^n F_n} \right) = \varphi + 0 = \varphi.
$$

Growing a golden rectangle

Consider an iterative procedure for constructing rectangles: given the nth rectangle in the sequence, construct the $(n + 1)$st as the minimal rectangle containing two copies of the nth, one horizontal and one vertical, side by side. If we begin with a 1×1 square, then the procedure generates a sequence of rectangles whose sides are Fibonacci numbers, as illustrated in Figure 2.5. Thus the rectangles in this sequence tend to the golden rectangle.

1×1   1×2   2×3   3×5   5×8

Figure 2.5.

If one begins the sequence with an arbitrary rectangle, the limit is always the golden rectangle [Walser, 2001].

Since $F_{n+1} = F_n + F_{n-1}$, we may write $F_{n+1}/F_n = 1 + 1/(F_n/F_{n-1})$ so that repeating the process yields, for $n \geq 2$ (and writing $F_2/F_1 = 1$),

$$
\frac{F_3}{F_2} = 1 + \frac{1}{1}, \quad \frac{F_4}{F_3} = 1 + \cfrac{1}{1 + \cfrac{1}{1}}, \quad \frac{F_5}{F_4} = 1 + \cfrac{1}{1 + \cfrac{1}{1 + \cfrac{1}{1}}},
$$

and so on to an expression for F_{n+1}/F_n that has $n-1$ "+" signs:

$$\frac{F_{n+1}}{F_n} = 1 + \cfrac{1}{1 + \cfrac{1}{\cdots \cfrac{1}{1 + \cfrac{1}{1}}}}.$$

Passing to the limit and employing $\lim_{n \to \infty} F_{n+1}/F_n = \varphi$ yields

$$\varphi = 1 + \cfrac{1}{1 + \cfrac{1}{1 + \cfrac{1}{1 + \cdots}}}.$$

Since $\sqrt{k + \sqrt{k + \sqrt{k + \sqrt{k + \cdots}}}} = (1 + \sqrt{1 + 4k})/2$ for $k \geq 1$ (see Challenge 2.16), we also have

$$\varphi = \sqrt{1 + \sqrt{1 + \sqrt{1 + \sqrt{1 + \cdots}}}}.$$

This limit expression for φ is illustrated in Figure 2.6, showing the convergence to φ of a sequence of nested radicals on the y-axis.

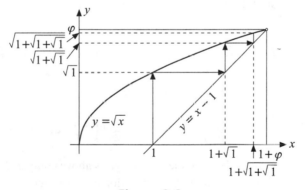

Figure 2.6.

2.4 π and the circle

The number π is usually defined as the ratio of the circumference to the diameter of a circle. So every student learns that a circle's circumference C

is given by $2\pi r$ and its area A by πr^2 where r denotes the radius of the circle. Why should the same constant π appear in both formulas?

An ancient and intuitive answer is given below in Figure 2.7. If we cut a circle of radius r in to a large number of congruent pie-shaped wedges, then we can arrange them in a figure that resembles a parallelogram with base πr (one-half the circumference) and height r, thus A should be πr^2.

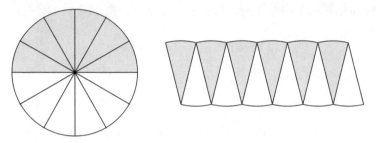

Figure 2.7.

Of course, that is not a proof, but it can be made rigorous with the use of limits. To do so, we circumscribe regular polygons about the circle, and note that the area of the polygon is equal to r times one-half its perimeter, and take the limit as the number of sides goes to infinity.

If we are going to use limits, we might as well use calculus, which we now do [Assmus, 1985]. Using definite integrals to express area and arc length, we have

$$A = 4 \int_0^r \sqrt{r^2 - x^2}\,dx \quad \text{and} \quad C = 4 \int_0^r r\,dx \Big/ \sqrt{r^2 - x^2}.$$

The change of variables $x = rt$ yields

$$A = \left(4 \int_0^1 \sqrt{1 - t^2}\,dt \right) r^2 \quad \text{and} \quad C = 2 \left(2 \int_0^1 dt \Big/ \sqrt{1 - t^2} \right) r.$$

So now the question is: How do the two integrals above compare? The identity $1 = (1 - x^2) + x^2$ followed by integration by parts yields

$$\int_0^1 \frac{1}{\sqrt{1 - t^2}}\,dt = 2 \int_0^1 \sqrt{1 - t^2}\,dt.$$

So if we set $2 \int_0^1 dt \Big/ \sqrt{1 - t^2} = \pi$ (the ratio of the circumference C to the diameter $2r$), we have $A = \pi r^2$.

2.5 The irrationality of π

It can be of no practical use to know that Pi is irrational, but if we can know, it surely would be intolerable not to know.

Edward Charles Titchmarsh

Theorem 2.4. π *is irrational.*

The following elegant proof [Niven, 1947] of the irrationality of π uses only elementary calculus. It is a classic.

Proof. Assume $\pi = a/b$ for positive integers a and b. For x in $[0, \pi] = [0, a/b]$, we define the polynomials

$$f(x) = \frac{b^n x^n (\pi - x)^n}{n!} = \frac{x^n (a - bx)^n}{n!},$$

$$F(x) = f(x) - f''(x) + f^{(4)}(x) - \cdots + (-1)^n f^{(2n)}(x),$$

the positive integer n to be specified later. Since $x^n (a - bx)^n$ has terms in x of degree at least n, $f(0)$ and the derivatives $f^{(k)}(0)$ for $0 \le k \le n - 1$ all equal 0; and since $f(x) = \frac{1}{n!} \sum_{k=n}^{2n} \binom{n}{k-n} a^{2n-k} (-b)^{k-n} x^k$, we have $f^{(k)}(0) = \frac{k!}{n!} \binom{n}{k-n} a^{2n-k} (-b)^{k-n}$ for $n \le k \le 2n$, which are all integers. The same is true at $x = \pi = a/b$ since $f(x) = f(\pi - x)$. Now we have

$$\frac{d}{dx} \left[F'(x) \sin x - F(x) \cos x \right] = F''(x) \sin x + F(x) \sin x = f(x) \sin x,$$

and so

$$\int_0^\pi f(x) \sin x \, dx = \left[F'(x) \sin x - F(x) \cos x \right]_0^\pi = F(\pi) + F(0). \quad (2.1)$$

Now $F(\pi) + F(0)$ is an integer since $f^{(k)}(\pi)$ and $f^{(k)}(0)$ are integers. But for x in $(0, \pi)$, $f(x) \sin x > 0$, the maximum value of $f(x)$ is $f(\pi/2)$, and hence

$$0 < f(x) \sin x \le f(\pi/2) = \frac{1}{n!} \left(\frac{\pi a}{4} \right)^n,$$

so that the integral in (2.1) is positive but arbitrarily close to 0 for n sufficiently large. Thus (2.1) is false, as is our assumption that π is rational. ∎

The Biblical value of π

We have all heard that somewhere in the Bible it says that π equals 3. The source of this claim is I Kings 7:23 (also II Chronicles 4:2), where the Bible (New International Version) describes a round "sea" (large bowl) made of metal, 10 cubits in diameter and 30 cubits in circumference.

The error in the approximation $\pi \approx 3$ is only about 4.5%. A circle of radius 1 has area π, and the inscribed regular dodecagon clearly has a slightly smaller area as seen in Figure 2.8. But if we cut a quarter of the dodecagon into nine pieces (three equilateral triangles and six isosceles triangles each with apex angle 150°) and rearrange them [Kürschak, 1898], then it is easy to see that the area of the inscribed dodecagon is exactly 3, so π is close to (but more than) 3.

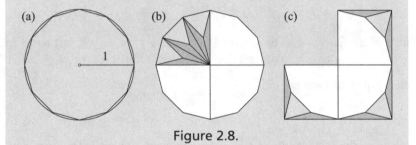

Figure 2.8.

One can also compare the circumference of the circle to the perimeter of the inscribed hexagon to obtain the same result.

2.6   The Comte de Buffon and his needle

In 1733, the French naturalist, historian, and mathematician Georges-Louis Leclerc, Comte de Buffon (1707–1788) posed the following problem: If we drop a needle at random onto a table ruled with equidistant parallel lines, what is the probability that the needle crosses one of them? This problem is now known as *Buffon's needle problem*, and its solution (adapted from [Grinstead and Snell, 1997]) provides a probabilistic method for approximating π.

Suppose the length of the needle is L, and the distance between each line on the table is D, where $0 < L \leq D$. See Figure 2.9a. We now have

Theorem 2.5. *A table is ruled with parallel lines D units apart, and a needle of length L is dropped onto the table ($0 < L \leq D$). The probability p that the needle falls crossing one of the lines is given by $p = 2L/\pi D$.*

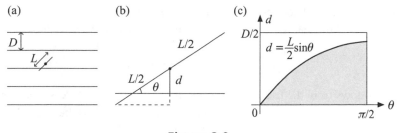

Figure 2.9.

Proof. Let d denote the distance from the center of the needle to the nearest line and θ the acute angle that the line of the needle makes with the parallel lines, where d is in $[0, D/2]$ and θ in $[0, \pi/2]$. See Figure 2.9b. Then the needle crosses a line if and only if $d < (L/2) \sin \theta$. Assuming that d and θ are uniformly distributed on their ranges, the probability p that the needle crosses a line is the ratio of the area of the shaded region in Figure 2.9c to the area $\pi D/4$ of the enclosing rectangle. The area of the shaded region is

$$\int_0^{\pi/2} \frac{L}{2} \sin \theta d\theta = \frac{L}{2},$$

and hence $p = (L/2)/(\pi D/4) = 2L/\pi D$. If we choose $L = D$, then the probability that the needle crosses one of the lines is $p = 2/\pi$. This leads to an experiment to approximate π: Drop a needle of unit length n times on a table with parallel lines 1 unit apart and count the number x of times the needle falls crossing a line, then $2n/x$ approximates π. However, as a practical method for obtaining a decimal approximation to π this experiment is useless. It would take over 10,000 tosses of the needle to obtain even the first decimal place of π with 95% confidence [Gridgeman, 1960].

The solution to Buffon's needle problem is an example of geometric probability and was a starting point for the field where geometry and probability meet—integral geometry. ∎

2.7 *e* as a limit

There are a variety of ways to define the number e. Perhaps the most common is as a limit: $e = \lim_{n \to \infty} (1 + 1/n)^n$. To justify this, we must show that the limit exists. This is usually done by showing that the sequence $\{(1 + 1/n)^n\}$ is bounded and monotone, since all bounded monotone sequences converge (a fact we will not prove here).

The following proof [Johnsonbaugh, 1974] is attractive because it has a visual element and does not use the natural logarithm.

We first show that for $0 \leq a < b$ and any positive integer n, we have

$$\frac{b^{n+1} - a^{n+1}}{b - a} < (n + 1)b^n. \tag{2.2}$$

This follows from the fact that the graph of $y = x^{n+1}$ is convex, so that the slope of a secant line joining two points on the graph is less than the slope of the tangent line at its right endpoint. See Figure 2.10.

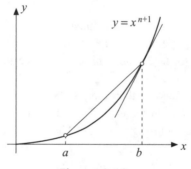

Figure 2.10.

If we clear fractions and rearrange terms, (2.2) becomes

$$b^n [(n + 1)a - nb] < a^{n+1}. \tag{2.3}$$

Setting $a = 1 + 1/(n + 1)$ and $b = 1 + 1/n$, the term in brackets in (2.3) reduces to 1 and we have

$$\left(1 + \frac{1}{n}\right)^n < \left(1 + \frac{1}{n + 1}\right)^{n+1}. \tag{2.4}$$

Setting $a = 1$ and $b = 1 + (1/2n)$, the term in brackets becomes $1/2$ so that

$$\left(1 + \frac{1}{2n}\right)^n \frac{1}{2} < 1.$$

Multiplying by 2 and squaring yields

$$\left(1 + \frac{1}{2n}\right)^{2n} < 4. \tag{2.5}$$

Inequalities (2.4) and (2.5) imply that the sequence $\{(1+1/n)^n\}$ is increasing and bounded above and hence converges.

Similarly we can show that the sequence $\{(1 + 1/n)^{n+1}\}$ is decreasing and bounded below (see Challenge 2.11) and that it has the same limit as the sequence $\{(1 + 1/n)^n\}$. Thus for any $n \geq 1$,

$$\left(1 + \frac{1}{n}\right)^n < e < \left(1 + \frac{1}{n}\right)^{n+1}. \tag{2.6}$$

This can be dramatically improved by replacing the exponents n and $n + 1$ by their geometric and arithmetic means $\sqrt{n(n+1)}$ and $n + 1/2$:

$$\left(1 + \frac{1}{n}\right)^{\sqrt{n(n+1)}} < e < \left(1 + \frac{1}{n}\right)^{n+\frac{1}{2}}. \tag{2.7}$$

To prove (2.7), we first show that if a and b are distinct positive numbers, then their *logarithmic mean* $(b - a)/(\ln b - \ln a)$ lies between the geometric mean $\sqrt{ab}$ and the arithmetic mean $(a + b)/2$. [That the geometric mean is always less than or equal to the arithmetic mean follows from expanding and simplifying $(\sqrt{a} - \sqrt{b})^2 \geq 0$.]

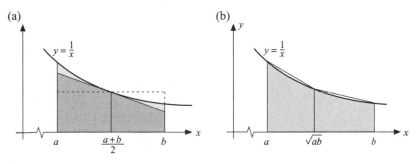

Figure 2.11.

Let $0 < a < b$. In Figure 2.11a we see that the area $\ln b - \ln a$ under the graph of $y = 1/x$ over $[a, b]$ is greater than its midpoint rule approximation $(b - a)(a + b)/2$, or equivalently, $(b - a)/(\ln b - \ln a) < (a + b)/2$. Figure 2.11b shows that the area $\ln b - \ln a$ is less than the sum $(b - a)/\sqrt{ab}$ of the areas of two trapezoids, or equivalently, $\sqrt{ab} < (b - a)/(\ln b - \ln a)$, which establishes the double inequality between the logarithmic mean and the geometric and arithmetic means: if $0 < a < b$, then $\sqrt{ab} < (b - a)/(\ln b - \ln a) < (a + b)/2$.

Now let $a = n$, $b = n + 1$ and take reciprocals to yield $2/(2n + 1) <$ $\ln[1 + (1/n)] < 1/\sqrt{n(n + 1)}$, which is equivalent to (2.7). So how does the double inequality (2.7) compare to (2.6)? For $n = 50$, (2.6) yields $2.69159 <$ $e < 2.74542$; whereas (2.7) yields $2.71824 < e < 2.71837$. The width of this interval is less than 0.25% of the width of the interval based on (2.6).

2.8   An infinite series for e

We now show that e can be expressed as another limit, and consequently as an infinite series. This limit is usually established using Taylor polynomials. We prove it using only integration by parts [Chamberland, 1999] to evaluate the sequence of partial sums of the series.

Theorem 2.6. $e = \displaystyle\lim_{n\to\infty} \left(1 + \dfrac{1}{1!} + \dfrac{1}{2!} + \cdots + \dfrac{1}{n!}\right) = 1 + \dfrac{1}{1!} + \dfrac{1}{2!} + \cdots.$

Proof. Let $a_n = \dfrac{1}{n!} \displaystyle\int_0^1 t^n e^{-t} dt$. Integration by parts yields $a_n = -\dfrac{1}{n!e} +$ a_{n-1} for any $n \geq 1$ and $a_0 = -\dfrac{1}{e} + 1$, so that

$$a_n = -\frac{1}{n!e} - \frac{1}{(n-1)!e} - \cdots - \frac{1}{1!e} - \frac{1}{e} + 1 = 1 - \frac{1}{e}\left(1 + \frac{1}{1!} + \frac{1}{2!} + \cdots + \frac{1}{n!}\right).$$

Since the integrand in a_n is bounded by 0 and 1, $0 \leq a_n \leq 1/n!$ and we have

$$0 = \lim_{n\to\infty} a_n = \lim_{n\to\infty} \left[1 - \frac{1}{e}\left(1 + \frac{1}{1!} + \frac{1}{2!} + \cdots + \frac{1}{n!}\right)\right],$$

which establishes the limit.   ■

2.9   The irrationality of e

We can now use the series for e to show that e is irrational. Our proof is from [Hardy and Wright, 1960].

Theorem 2.7. *e is irrational.*

Proof. Assume e is rational, with $e = p/q$ for positive integers p and q. Then

$$\frac{p}{q} = e = \sum_{n=0}^{q} \frac{1}{n!} + \sum_{n=q+1}^{\infty} \frac{1}{n!}.$$

Transpose the first sum on the right to the left, and multiply each side by $q!$:

$$p(q-1)! - \sum_{n=0}^{q} \frac{q!}{n!}$$
$$= \frac{1}{q+1} + \frac{1}{(q+1)(q+2)} + \frac{1}{(q+1)(q+2)(q+3)} + \cdots$$
$$< \frac{1}{q+1} + \frac{1}{(q+1)^2} + \frac{1}{(q+1)^3} + \cdots = \frac{1}{q}. \tag{2.8}$$

Thus the expression on the left in (2.8) is an integer, while the expression on the right in (2.8) lies strictly between 0 and 1. Hence our assumption that $e = p/q$ is false, so e must be irrational. ∎

2.10 Steiner's problem on the number e

In Heinrich Dörrie's classic work *100 Great Problems of Elementary Mathematics* [Dörrie, 1965] eight of the 100 problems are attributed to Jakob Steiner (1796–1863). This is problem 89:

For what positive x is the x-th root of x the greatest?

A common approach is to use calculus to find the maximum value of the function $f(x) = x^{1/x}$. Here is a simple solution that uses only the convexity of the exponential function and the monotonicity of the power function.

Solution. We answer by showing that for x positive, $x^{1/x} \leq e^{1/e}$. See Figure 2.12.

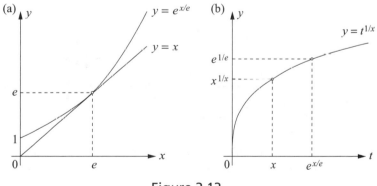

Figure 2.12.

In Figure 2.12a we have drawn the line $y = x$ tangent to the curve $y = e^{x/e}$ at (e,e). Thus for $x > 0$, $x \leq e^{x/e}$. Raising each side to the $1/x$ power (see Figure 2.12b for the case $x > 1$) yields the desired result (the other case differs only in concavity). In Section 12.4 we use this inequality to establish the arithmetic mean-geometric mean inequality for n numbers.

2.11   The Euler-Mascheroni constant

The Euler-Mascheroni constant γ measures how the partial sums of the divergent harmonic series differ from the natural logarithm function. It is defined by

$$\gamma = \lim_{n \to \infty} \left(\sum\nolimits_{k=1}^{n} (1/k) - \ln(n+1) \right). \tag{2.9}$$

To show that (2.9) makes sense, i.e., that the limit exists, we consider the sequence $\{a_n\}$ given by

$$a_n = \sum\nolimits_{k=1}^{n} (1/k) - \ln(n+1) \tag{2.10}$$

for $n \geq 1$, and show that it is increasing and bounded from above. Now

$$a_{n+1} - a_n = \frac{1}{n+1} - \ln(n+2) + \ln(n+1) = \frac{1}{n+1} - \int_{n+1}^{n+2} \frac{1}{x} \, dx > 0,$$

as illustrated in Figure 2.13, where the region shaded gray represents the difference between the area $1/(n+1)$ of a rectangle and the area under the graph of $y = 1/x$ over the interval $[n+1, n+2]$.

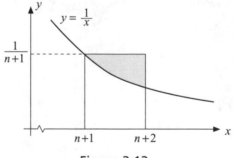

Figure 2.13.

Furthermore, $a_n < 1$ for all n, as shown in Figure 2.14, where a_n is represented by the n gray shaded regions above the graph of $y = 1/x$ over the interval $[1, n+1]$.

Hence the limit in (2.9) exists, and evaluated to twenty decimal places, $\gamma \approx 0.57721566490153286060$. It is still unknown whether γ is rational or irrational.

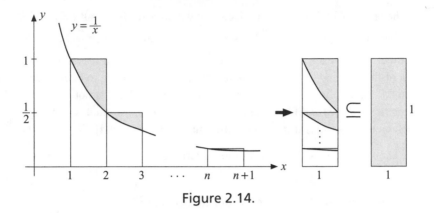

Figure 2.14.

2.12 Exponents, rational and irrational

If one knows whether or not two numbers a and b are rational or irrational, it is easy to determine whether $a + b$ and ab are rational or irrational. For example, if a and b are both irrational, then $a + b$ and ab may be either rational or irrational, whereas if one of a and b is rational and the other is irrational, then both $a + b$ and ab are irrational.

What can one say about a^b knowing whether a and b are rational or irrational? If both a and b are rational, then a^b may be either rational or irrational, for example $2^{1/2} = \sqrt{2}$, $4^{1/2} = 2$. As we shall now see, the same is true in the cases where at least one of a and b is irrational [Jones and Toporowski, 1973].

Theorem 2.8. *An irrational number to an irrational power may be rational.*

Proof. To prove this theorem, we need only give an example a^b where a and b are irrational and a^b rational. If $\sqrt{2}^{\sqrt{2}}$ is rational, then it is our example. If $\sqrt{2}^{\sqrt{2}}$ is irrational, then $(\sqrt{2}^{\sqrt{2}})^{\sqrt{2}} = 2$ is our example. ∎

Theorem 2.9. *An irrational number to an irrational power may be irrational.*

Proof. If $\sqrt{2}^{\sqrt{2}}$ is irrational, then it is our example. If $\sqrt{2}^{\sqrt{2}}$ is rational, then $\sqrt{2}^{\sqrt{2}+1} = \sqrt{2}^{\sqrt{2}} \cdot \sqrt{2}$ is our example. ∎

The cases where we have a rational to an irrational power or an irrational to a rational power are examined in Challenge 2.14.

The astute reader will have noticed that we proved Theorems 2.8 and 2.9 without knowing whether $\sqrt{2}^{\sqrt{2}}$ is rational or irrational. We used only the fact that $\sqrt{2}$ is irrational. In fact, $\sqrt{2}^{\sqrt{2}}$ is irrational, since it is the square root of the *Gelfand-Schneider number* $2^{\sqrt{2}}$, which is known to be transcendental [a (possibly complex) number is *transcendental* if it is not algebraic; a number is *algebraic* if it is the root of a non-zero polynomial with integer coefficients; all (real) transcendental numbers are irrational]. However, no one knows whether $\sqrt{2}^{\sqrt{2}^{\sqrt{2}}}$ is rational or irrational.

Hilbert and the Gelfand-Schneider number $2^{\sqrt{2}}$

On August 8, 1900 David Hilbert (1862–1943) presented an address at the Second International Congress of Mathematicians in Paris. In this address Hilbert presented twenty-three problems (in the printed version, but only ten in the oral presentation) the solutions to which he believed would profoundly impact the course of mathematical research in the twentieth century.

In his seventh problem, Hilbert conjectured that "the expression a^b, for an algebraic base a [not equal to 0 or 1] and an irrational algebraic exponent b, e.g. $2^{\sqrt{2}}$ or $e^\pi = i^{-2i}$, always represents a transcendental or at least an irrational number." The conjecture was proved by A. O. Gelfand and T. Schneider independently in 1934. However, it is still not known whether a^b is transcendental when both a and b are transcendental.

2.13   Challenges

2.1 Create visual proofs similar to the one in Figure 2.1 to show that (a) $\sqrt{3}$ and (b) $\sqrt{5}$ are irrational.

2.2 Prove that $\sqrt{2}$ is irrational by showing that there are no solutions to the congruence $m^2 \equiv 2n^2 \pmod 3$ with m and n relatively prime.

2.3 Prove that $\sqrt[n]{2}$ is irrational for all $n \geq 3$. [Hint: use *Fermat's Last Theorem*: $x^n + y^n = z^n$ has no solutions in positive integers for $n \geq 3$.]

2.4 Prove that the length of the diagonal of a regular pentagon with side length 1 is the golden ratio φ.

2.5 An equilateral triangle with side 1 is partitioned into three equilateral triangles and a trapezoid by line segments parallel to the sides, as shown in Figure 2.15a. Show that the shaded trapezoids in Figures 2.15b and 2.15c are similar if and only if $x = 1/\varphi$.

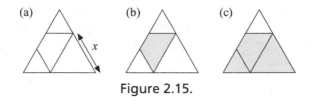

Figure 2.15.

2.6 Let A denote the area of an annulus of outer radius a and inner radius b and E the area of an ellipse with semi-major and semi-minor axes of lengths a and b, respectively, as illustrated in Figure 2.16. Find the ratio of a to b if A and E are equal [Rawlins, 1995].

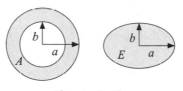

Figure 2.16.

2.7 Three identical rectangles measuring $a \times b$ with $a < b$ fit together as shown in Figure 2.17a, each perpendicular to the other two and intersecting at their centers. Their vertices determine a convex polyhedron with triangular faces as shown in Figure 2.17b. Show that the polyhedron is a regular icosahedron if and only if $b/a = \varphi$, i.e., only golden rectangles are sections of a regular icosahedron.

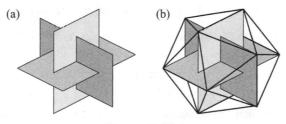

Figure 2.17.

2.8 Let $[[x]]$ denote the *nearest integer function*, i.e., the integer nearest to x (when x is a half-integer, it is the nearest even integer). Show that for $n \geq 1$,

$$F_n = [[\varphi^n / \sqrt{5}]].$$

2.9 A sequence of the form $\{a, b, a+b, a+2b, 2a+3b, \cdots\}$ with $ab \neq 0$ in which each term after the second is the sum of the preceding two is called a *generalized Fibonacci sequence*. Are any of them geometric progressions?

2.10 Prove that $\pi < 22/7$. [Hint: Prove $22/7 - \pi = \int_0^1 [x^4(1-x)^4/(1+x^2)]dx$. This was Problem A-1 on the 29th William Lowell Putnam Mathematical Competition in 1968.]

2.11 (a) Show that for $0 \leq a < b$ and any positive integer n, we have

$$(n+1)a^n < \frac{b^{n+1} - a^{n+1}}{b - a}.$$

   (b) Use (a) to show that the sequence $\{(1+1/n)^{n+1}\}$ is decreasing and bounded below.

   (c) Show that the sequences $\{(1+1/n)^n\}$ and $\{(1+1/n)^{n+1}\}$ have the same limit, and consequently $(1+1/n)^n < e < (1+1/n)^{n+1}$ for all $n \geq 1$.

2.12 Show that $\lim\limits_{n \to \infty} \dfrac{\sqrt[n]{n!}}{n} = \dfrac{1}{e}$.

2.13 Which is larger, e^π or π^e?

2.14 (a) Show that a rational number to an irrational power may be irrational.

   (b) Show that a rational number to an irrational power may be rational.

   (c) Why do we not ask about an irrational number to a rational power?

2.15 If $\{a_n\}$ is as in (2.10), show that $\gamma - a_n > 1/2(n+1)$. As a consequence, it converges very slowly to γ.

2.16 Show that for $k \geq 2$, $\sqrt{k + \sqrt{k + \sqrt{k + \sqrt{k + \cdots}}}} = (1 + \sqrt{1+4k})/2$. [Hint: Consider the sequence $\{x_n\}$ defined by $x_1 = \sqrt{k}$ and $x_{n+1} = \sqrt{k + x_n}$ and show that it is increasing and bounded above.]

CHAPTER 3

Points in the Plane

Mighty is geometry; joined with art, resistless.
Euripedes

Geometry is the art of correct reasoning on incorrect figures.
George Pólya

In this chapter we present some intriguing results, and their delightful proofs, about some of the simplest geometric configurations in the plane. These include figures consisting solely of points and lines, including those constructed from the lattice points in the plane. We will deal with structures such as triangles, quadrilaterals, and circles in later chapters.

3.1 Pick's theorem

Pick's theorem is admired for its elegance and its simplicity; it is a gem of elementary geometry. Although it was first published in 1899, it did not attract much attention until seventy years later when Hugo Steinhaus included it in the first edition of his lovely book *Mathematical Snapshots* [Steinhaus, 1969]. Georg Alexander Pick (1859–1942) was born in Vienna but lived much of his life in Prague. Pick wrote many mathematical papers in the areas of differential equations, complex analysis, and differential geometry. Sadly, Pick was arrested by the Nazis in 1942 and sent to the concentration camp at Theresienstadt, where he perished.

A *lattice point* in the plane is a point with integer coordinates, and a *lattice polygon* is a polygon whose vertices are lattice points. A polygon is *simple* if it has no self-intersections. Pick's theorem gives the area $A(S)$ of a simple lattice polygon S in terms of the number i of interior lattice points and the number b of lattice points on the boundary: $A(S) = i + b/2 - 1$. For example, for the lattice polygon S_2 in Figure 3.1, $i = 15$, $b = 8$, and so $A(S_2) = 18$.

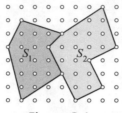

Figure 3.1.

Today elementary school students often study Pick's theorem. Using a geoboard (a wooden board with a rectangular array of nails driven part way into the board) and rubber bands to form polygons, students can easily calculate areas.

There are a variety of ways to prove Pick's theorem. Our proof [Varberg, 1985] is elegant in its simplicity; it is direct and intuitive. First note that any lattice polygon can be partitioned into a union of lattice triangles (this can be easily proved by induction on the number of sides, using an interior diagonal of the polygon).

We begin by defining a *visibility angle* θ_k (in degrees) for each lattice point P_k of the polygon, the angle with which one can "see" into the polygon and a weight $w_k = \theta_k/360°$. For example, $w_k = 1$ for an interior lattice point, $w_k = 1/2$ for a boundary point that is not a vertex, and $w_k = 1/4$ for a right-angled vertex. Now let $W(S) = \sum_{P_k \in S} w_k$ denote the total weight of the polygon S. Surprisingly, we have

Lemma 3.1. *If S is a simple lattice polygon, then $W(S) = A(S)$.*

Proof. Like area, the weight function W is *additive*: if S_1 and S_2 are disjoint (except for perhaps a shared boundary) and $S = S_1 \cup S_2$ (as in Figure 3.1), then $W(S) = W(S_1) + W(S_2)$. This follows from the observation that the visibility angles in S_1 and S_2 at a common lattice point add to give the visibility angle in S for that point. We now examine several cases.

Consider Case 1, a lattice rectangle whose sides are horizontal and vertical, as illustrated in Figure 3.2a. Each lattice point corresponds to a square or rectangle whose area equals the weight of the lattice point, so the weight and area of the rectangle are equal.

In Figure 3.2b we have Case 2, a right triangle S with legs horizontal and vertical, and $W(S) = A(S)$ follows by division by 2 from Case 1.

In Case 3, Figures 3.2c and 3.2d, we surround a general lattice triangle S by Case 2 triangles and Case 1 rectangles, and again $W(S) = A(S)$ by the additivity of A and W. To complete the proof we decompose an arbitrary simple lattice polygon into a union of lattice triangles and use the additivity of W. ∎

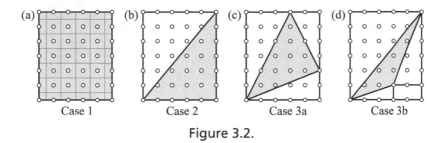

Figure 3.2.

We now prove

Pick's Theorem 3.1. *For a simple lattice polygon S with i interior lattice points and b boundary lattice points, $A(S) = i + b/2 - 1$.*

Proof. If a simple polygon has n vertices, then the sum of the sizes of the n interior vertex angles is $(n - 2)180°$ (for a proof see Section 4.1). Consequently, the sum of the visibility angles at all points on the boundary of S is $(b - 2)180°$. If we let I and B denote the interior and boundary of S, respectively, then

$$A(S) = W(S) = \sum_{P_k \in I} w_k + \sum_{P_k \in B} w_k$$

$$= i + \frac{(b - 2)180°}{360°} = i + \frac{b}{2} - 1. \qquad ∎$$

3.2 Circles and sums of two squares

The problem of representing a number as a sum of squares has a long history, going back at least to the time of Pythagoras. Finding right triangles with integer sides amounts to finding positive integers a, b, and c such that c^2 is a sum of two squares, $a^2 + b^2$. More generally we can ask which integers n can be represented as a sum of two squares of integers and how many representation there are.

Let $r_2(n)$ denote the number of ways we can represent the integer n as a sum of two squares (positive, negative, or zero). That is, $r_2(n)$ denotes the number of solutions in integers (x, y) to the equation $x^2 + y^2 = n$. For example, $r_2(5) = 8$ since the solutions to $x^2 + y^2 = 5$ are $(1,2), (2,1), (-1,2), (2,-1), (1,-2), (-2,1), (-1,-2),$ and $(-2,-1)$. Because $r_2(n) = 0$ whenever n has the form $4k - 1$, $r_2(n)$ is a very erratic function (See Challenge 3.6).

But we can ask: What is the average value of $r_2(k)$ for $1 \le k \le n$? We define $N_2(n)$ to be the number of solutions in integers to $x^2 + y^2 \le n$, and then the average of $r_2(k)$ for $1 \le k \le n$ is

$$\frac{r_2(1) + r_2(2) + \cdots + r_2(n)}{n} = \frac{N_2(n)}{n}.$$

Computation of $N_2(n)$ and $N_2(n)/n$ yields the table

n	1	2	3	4	5	10	20	50	100
$N_2(n)$	5	9	9	13	21	37	69	161	317
$N_2(n)$	5	4.5	3	3.25	4.2	3.7	3.45	3.22	3.17

The average value of r_2 over all the integers is the limit of $N_2(n)/n$ as $n \to \infty$ if the limit exists. It does, and we have

Theorem 3.2. $\lim_{n\to\infty} \dfrac{N_2(n)}{n} = \pi$.

Proof. The proof is based on a geometric interpretation of $N_2(n)$, and is due to Carl Friedrich Gauss (1777–1855): $N_2(n)$ is the number of points in or on a circle $x^2 + y^2 = n$ with integer coordinates. For example, $N_2(10) = 37$ since the circle centered at the origin of radius $\sqrt{10}$ contains 37 lattice points, as illustrated in Figure 3.3a. If we draw a square with area 1 centered at each of the 37 points, then the total area of the squares (in gray) is also $N_2(10)$. Thus we would expect the area of the squares to be approximately the area of the circle, or in general, $N_2(n)$ to be approximately $\pi(\sqrt{n})^2 = \pi n$.

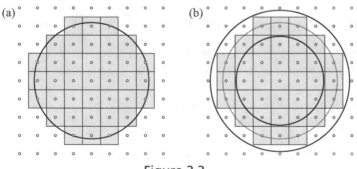

Figure 3.3.

If we expand the circle of radius $\sqrt{n}$ by half the length of the diagonal ($\sqrt{2}/2$) of a square of area 1, then the expanded circle contains all the squares. If we contract the circle by the same amount, then the contracted circle would be contained in the union of all the squares (see Figure 3.3b). Thus

$$\pi(\sqrt{n} - \sqrt{2}/2)^2 < N_2(n) < \pi(\sqrt{n} + \sqrt{2}/2)^2.$$

Dividing each term by n and applying the squeeze theorem for limits yields the desired result. ∎

The same procedure can be extended to higher dimensions. See Challenge 3.7.

3.3 The Sylvester-Gallai theorem

In 1893 James Joseph Sylvester (1814–1897) proposed the following problem in the Mathematical Questions column of the *Educational Times*:

> Prove that it is not possible to arrange any finite number of real points so that a right line through every two of them shall pass through a third, unless they all lie in the same right line.

By "right line" Sylvester meant a straight line. Over forty years later Tibor Gallai (1912–1992) solved the problem, prior to the appearance of the following *American Mathematical Monthly* problem proposed by Paul Erdős [Erdős, 1943]:

> Let n given points have the property that the straight line joining any two of them passes through a third point of the set. Show that the n points lie on a straight line.

Erdős was apparently unaware of Sylvester's problem proposal and Gallai's solution. In the years since, several additional proofs of what is now known as the *Sylvester-Gallai theorem* have appeared. The simple, direct, and elegant one we present is by L. M. Kelly [Coxeter, 1948].

The Sylvester-Gallai Theorem 3.3. *If a finite set of points in the plane are not all on one line then there is a line through exactly two of the points.*

Proof. Let P denote the set of points, and L the set of lines determined by the points p in L. For each p in P and ℓ in L with the property that p is not on ℓ, we now consider the pair (p, ℓ) and compute the distance from p to ℓ. Since both P and L are finite, there are a finite number of pairs (p, ℓ)

to consider, and hence we can find the pair (p_0, ℓ_0) for which the distance is a minimum. We now show that ℓ_0 contains exactly two points of P.

Suppose there are three or more points of P on ℓ_0. Let q (which may or may not be a member of P) be the foot of the perpendicular from p_0 to ℓ_0. Of the points in P on ℓ_0, at least two of them must be on the same side of q, call them p_1 and p_2, with p_1 (which may coincide with q) between q and p_2. See Figure 3.4. Now let ℓ_1 be the line in L determined by p_0 and p_2, and note that the distance from p_1 to ℓ_1 is less than the distance from p_0 to ℓ_0. But this cannot be, since the pair (p_0, ℓ_0) was chosen as the pair with minimum distance. Hence ℓ_0 contains exactly two points of P. ■

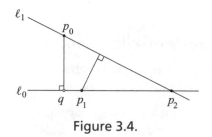

Figure 3.4.

3.4   Bisecting a set of 100,000 points

Suppose 100,000 points are selected at random inside a circle. Is it always possible to draw a line across the circle missing every point so that exactly 50,000 points lie on each side of the line?

The answer is yes [Gardner, 1989]. In fact, the answer is yes for *any* even number $2n$ of points inside the circle. A procedure for finding such a line is illustrated in Figure 3.5 for six points, but it applies to any even number.

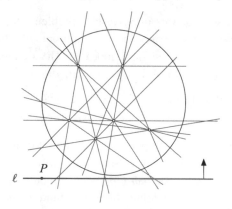

Figure 3.5.

Consider all the lines determined by pairs of points inside the circle, choose a point P outside the circle not lying on any of the lines, and draw a line ℓ through P not intersecting the circle, as shown in Figure 3.5. Now begin to rotate ℓ about P. As we do so, ℓ will pass over each of the points one at a time because P does not lie on any of the lines passing through two or more points. After ℓ has passed over n points, exactly n of the $2n$ points will lie on each side of the line.

3.5   Pigeons and pigeonholes

The *pigeonhole principle*, also known as *Dirichlet's box* or *drawer principle*, is a simple but powerful tool for solving problems and proving theorems. It was first stated by Peter Gustav Lejeune Dirichlet (1805–1859) in 1834. It is based on the elementary observation that if m pigeons fly into n pigeonholes with $m > n$, then at least one pigeonhole will have more than one pigeon. For example, in Figure 3.6 we see $m = 10$ pigeons in $n = 9$ pigeonholes.

Figure 3.6.

Ross Honsberger [Honsberger, 1973] relates the following story about Paul Erdős and Louis Pósa. When Pósa was about eleven years old, Erdős asked him the following question while having lunch one day: "Prove that if you have $n + 1$ positive integers less than or equal to $2n$, some pair of them are relatively prime." Pósa responded about a half a minute later: "If you have $n + 1$ positive integers less than or equal to $2n$, two of them must be consecutive and thus relatively prime."

As noted in [Schattschneider, 2006], this is the pigeonhole principle in its purest form: Since "there are a maximum of n nonconsecutive 'pigeonholes' among $2n$ of them in a line, and so ... placing $n + 1$ numbers in the $2n$ slots would force (at least) two of them to be in consecutive holes." For a related problem concerning the first $2n$ positive integers, see Challenge 3.8.

Here are some additional examples of proofs based on this principle, each concerning points in the plane or on the surface of a sphere.

Example 3.1. In any set of five points inside an equilateral triangle of side length 1, there exists a pair of points no more than $1/2$ unit apart.

Divide the triangle into four congruent equilateral triangles (the pigeonholes) of side length $1/2$. By the pigeonhole principle, at least two of the five points (the pigeons) will be in the same small triangle.

Example 3.2. (Problem A2 on the 63rd William Lowell Putnam Mathematical Competition, 2002). Given any five points on a sphere, show that some four of them must lie on a closed hemisphere.

Draw a great circle through two of the points. There are two closed hemispheres with it as boundary, and each of the other three points lies in one of them. By the pigeonhole principle, two of the three points lie in the same hemisphere, and that hemisphere thus contains four of the five given points.

More generally, if m pigeons fly into n pigeonholes with $m > n$, then at least one pigeonhole will have at least $\lceil m/n \rceil$ pigeons, where $\lceil x \rceil$ denotes the ceiling function, the smallest integer greater than or equal to x.

Example 3.3. In any set of 51 points inside a square with side length 1, there is always a subset of three points that can be covered by a circular disk of radius $1/7$.

Divide the square into 25 congruent subsquares of side length $1/5$. By the pigeonhole principle, there must be one subsquare with at least $\lceil 51/25 \rceil = 3$ points. But each subsquare of side $1/5$ fits inside a circle of diameter $2/7$ since $\sqrt{2}/5 < 2/7$.

As is often the case, a principle from discrete mathematics can be extended and applied in the continuous case. This is true for the pigeonhole principle. The primary difficulty often lies in deciding what are the pigeons and what are the pigeonholes. Counting objects is frequently replaced by measuring them. Here is one example [Strzelecki and Schenitzer, 1999].

Example 3.4. If you paint more than one-half of a sphere, then both endpoints of one of its diameters must be painted.

Suppose not. Reflecting the sphere in its center, the image of every painted point is an unpainted point. Thus the image N of the set P of painted points is the set of unpainted points. Hence N and P are disjoint subsets of the sphere with equal areas. This is impossible since the area of P is more than one-half the area of the sphere.

3.6 Assigning numbers to points in the plane

Each year in April, approximately 500 high school students are invited to take the USAMO (United States of America Mathematical Olympiad). The USAMO is a six question two day, nine hour essay/proof examination. The purpose of the USAMO is to "identify and encourage the most creative secondary mathematics students in the country [and] to indicate the talent of those who may become leaders in the mathematical sciences of the next generation." While all problems can be solved with pre-calculus mathematics, the problems would challenge most professional mathematicians. Here is Problem 6 from the 2001 USAMO and a solution by one of the contestants [Andreescu and Feng, 2002]:

> *Problem* 6. Each point in the plane is assigned a real number such that, for any triangle, the number at the center of its inscribed circle is equal to the arithmetic mean of the three numbers at its vertices. Prove that all points in the plane are assigned the same number.

Solution (Michael Hamburg, St. Joseph's High School, South Bend, Indiana). A lowercase letter denotes the number assigned to the point labeled with the corresponding uppercase letter. Let A, B be arbitrary distinct points, and consider a regular hexagon $ABCDEF$ in the plane (see Figure 3.7). Let lines CD and FE intersect at G. Let ℓ be the line through G perpendicular to line ED. Then A, F, E and B, C, D are symmetric to each other, respectively, with respect to line ℓ. Hence triangles CEG and DFG share the

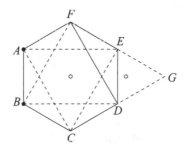

Figure 3.7.

same incenter, i.e., $c + e = d + f$, and triangles ACE and BDF share the same incenter, i.e., $a + c + e = b + d + f$. Therefore $a = b$ and we are done.

Each year the Clay Mathematics Institute gives the Clay Olympiad Scholar Award to the solution judged most creative. Hamburg received the award for this solution in 2001. In their press release, the Clay Mathematics Institute wrote

> The twelve judges of the USA Mathematical Olympiad Competition recommended Mr. Hamburg, an eleventh grade student at Saint Joseph's High School, as the contestant whose correct solution to a mathematics problem in the competition best fulfilled the CMI selection criteria of elegance, beauty, imagination, and depth of insight. They unanimously cited Mr. Hamburg for his solution to Problem 6 on the test, which only 9 of the 270 final round contestants answered correctly. Michael invented a particularly ingenious construction and then concisely and elegantly derived the result. The judges declared that the diagram in his solution was virtually a "proof without words."

3.7   Challenges

3.1 Is it possible to construct an equilateral lattice triangle?

3.2 Does a version of Pick's theorem hold for three-dimensional lattice polyhedra? [Hint: consider a tetrahedron with vertices $(0,0,0)$, $(1,0,0)$, $(0,1,0)$, and $(1,1,k)$ for a positive integer k.]

3.3 Why does the Sylvester-Gallai theorem fail to hold for infinite sets of points?

3.4 Given $n \geq 3$ non-collinear points in the plane, the Sylvester-Gallai theorem tells us that there must be at least one line containing exactly two points. There may be more than one such line. In fact, Kelly and Moser [1958] have proved that there must be at least $3n/7$ lines with exactly two points. Show that there exists an arrangement of n points in the plane such that exactly $3n/7$ lines contain exactly two points (and hence the lower bound of $3n/7$ cannot be improved). [Hint: let $n = 7$.]

3.5 Consider a rectangular grid of lattice points with m points in each of n rows. Find the number of rectangles with vertices at lattice points and sides parallel to the sides of the grid.

3.6 Prove that if $n = 4k - 1$ for a positive integer k, then $r_2(n) = 0$.

3.7 For any positive integer n, let $N_3(n)$ be the number of triplets (x, y, z) of integers such that $x^2 + y^2 + z^2 \leq n$. Show that $N_3(n)$ is approximately $4\pi n \sqrt{n}/3$ for large values of n.

3.8 Each of the vertices of a regular pentagon is colored either black or white. Both colors are used. Prove that there are three vertices of the pentagon that are colored the same color, and that they form an isosceles triangle.

3.9 Prove that if you select $n + 1$ of the first $2n$ positive integers, then some two of the selected integers have the property that one divides the other.

3.10 (a) Every point in the plane is painted one of two colors. Prove that there exist two points of the same color exactly one inch apart.

   (b) Every point in the plane is painted one of three colors. Prove that there exist two points of the same color exactly one inch apart.

   (c) Is part (b) true if "three" is replaced by "nine"?

3.11 Every point in the plane is painted red or blue. Prove that some rectangle has its vertices all the same color.

CHAPTER 4

The Polygonal Playground

> *Next above them come the Nobility, of whom there are several degrees, beginning with Six-sided Figures, or Hexagons, and from thence rising in the number of their sides till they receive the honourable title of Polygonal, or many-sided.*
>
> Edwin A. Abbott, *Flatland*

Polygons, along with lines and circles, constitute the earliest collection of geometric figures studied by humans. Polygons, along with their associated star polygons, appear frequently throughout history in symbolic roles, often in religious and mystical settings. Book IV of the *Elements* of Euclid deals exclusively with the construction of certain regular polygons (triangles, squares, pentagons, hexagons and 15-gons) and how to inscribe and circumscribe them in and about circles.

In this chapter we consider some remarkable results and their proofs that apply to general polygons. In subsequent chapters we study particular polygons such as triangles, squares, general quadrilaterals, polygons that tile the plane, etc.

4.1　Polygonal combinatorics

By *polygon*, we mean a closed figure with a finite number n of sides ($n \geq 3$), each of which is a line segment. Polygons can be convex, concave, and star. See Figure 4.1. We use the word *n-gon* to indicate a polygon with n sides (and angles).

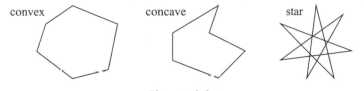

Figure 4.1.

Theorem 4.1. *In a convex n-gon, we have:*

  (i) *the sum of the angle measures is* $(n-2)180°$

  (ii) *there are* $n(n-3)/2$ *diagonals*

  (iii) *the diagonals intersect in at most* $\dbinom{n}{4}$ *interior points.*

Proof. Since the n-gon is convex, we can select one vertex and draw $n-3$ diagonals connecting it to all the other vertices except the two neighboring vertices. The $n-3$ diagonals divide the n-gon into $n-2$ triangles, so the sum of all the interior angles is $(n-2)180°$. For (ii), we note that $n-3$ diagonals terminate at each vertex, so the total number of endpoints of diagonals is $n(n-3)$. Since each diagonal has two endpoints, there are $n(n-3)/2$ diagonals. For (iii), each interior point of intersection of two diagonals is also the point of intersection of the diagonals of at least one quadrilateral whose four vertices are vertices of the n-gon (see Figure 4.2), and four vertices of the n-gon can be chosen in $\dbinom{n}{4}$ ways. ∎

Figure 4.2.

Now consider an n-gon and all of its diagonals. How many triangles can be found in such a figure? We answer the question in the case where no three diagonals meet at an interior point, as in Figure 4.2.

Theorem 4.2. *Let P be a convex n-gon with the property that no three diagonals meet at an interior point. Then the number of triangles in P whose vertices are either interior points or vertices of P is*

$$\binom{n}{3} + 4\binom{n}{4} + 5\binom{n}{5} + \binom{n}{6}.$$

Proof [Conway and Guy, 1996]. We count the triangles according to the number of vertices that each triangle shares with P. There are $\binom{n}{3}$ triangles with all three vertices in common with P (see Figure 4.3a), $4\binom{n}{4}$ triangles that have exactly two vertices in common with P (see Figure 4.3b), $5\binom{n}{5}$ triangles that have exactly one vertex in common with P (see Figure 4.3c), and finally $\binom{n}{6}$ triangles all of whose vertices are interior points of P. ∎

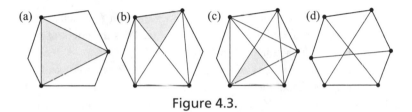

Figure 4.3.

A similar question is the following: into how many regions do the diagonals of a polygon partition the interior of the polygon?

Theorem 4.3. *Let P be a convex n-gon with the property that no three diagonals meet at an interior point. Then the number of regions in the resulting partition of P is*

$$\binom{n}{4} + \binom{n-1}{2}.$$

Proof [Honsberger, 1973, Freeman, 1976]. In this proof we use the moving line argument we introduced in Section 3.4. Since P has only a finite number of sides and diagonals, we may select a point Q outside of P such that Q does not lie on any of the lines formed by extending the sides and diagonals of P. See Figure 4.4.

Draw a line ℓ through Q not intersecting P, and begin rotating ℓ about Q. We count the number of regions in P by counting a particular region when the line ℓ first intersects it as ℓ rotates about Q. A region will be first encountered by ℓ when ℓ passes through either a vertex or an interior point of P (i.e., a point where two diagonals intersect within P).

Consider an interior point I of P. Before ℓ passes through I, ℓ will have intersected three of the four regions having I as a vertex, and these regions

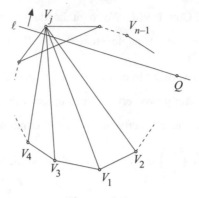

Figure 4.4.

will have already been counted. So as ℓ passes through I, exactly one new region is encountered, and it counted at this time. Thus the count of regions increases by one for each interior point of P.

Now consider the vertices of P. Each vertex of P is a vertex of $n-2$ triangular regions. Label the vertices in the order they are encountered by ℓ as it rotates about Q: $V_1, V_2, \ldots, V_n$. At V_1 the count of regions increases by $n-2$, at V_2 the count increases by $n-3$ since one triangle (the one which also has a vertex at V_1) has already been counted, and so on, until ℓ passes through V_{n-1} and V_n where it does not encounter any new triangles that have not already been counted. To be precise, when ℓ passes through V_j, each of the line segments $V_1V_j, V_2V_j, \ldots, V_{j-1}V_j$ intersects the trailing half-plane of ℓ and is associated with a triangle that has already been counted.

Since we obtain one new region for each interior point of P and $n-1-j$ regions for vertex V_j (for $j = 1, 2, \ldots, n-1$, and 0 for V_n), and since ℓ passes through each region just once, the number of regions in P is

$$\binom{n}{4} + (n-2) + (n-3) + \cdots + 2 + 1 + 0 + 0 = \binom{n}{4} + \binom{n-1}{2}.$$

∎

4.2 Drawing an *n*-gon with given side lengths

If a, b, c are lengths of the sides of a triangle, then the triangle is easy to draw. What if we are given numbers $a_1, a_2, \ldots, a_n$ which are the sides of an n-gon in that order. Can we draw the figure even if we have no knowledge of the measures of the angles? The surprising answer is: Yes.

As you might suspect, we will need more than the tools used by the Greeks for the classical constructions, the compass and straightedge. We also need a large piece of cardboard or heavy paper, and scissors. Draw an arc of a large circle on the cardboard, mark a point near one end of it, and construct chords of lengths $a_1, a_2, \ldots, a_n$ (in that order) on it, so that we have a polygonal line inscribed in the arc. Using the straightedge draw a radius from the center to each vertex of the polygonal line. Use scissors to cut out the figure, and fold it on each of the radii to form the sides of a pyramid whose base is the desired n-gon with side lengths $a_1, a_2, \ldots, a_n$.

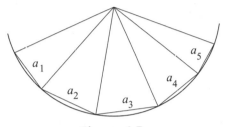

Figure 4.5.

Patience and the 65537-gon

The first regular polygons to be inscribed in a circle (by means of a compass and straightedge) are described in Book IV of the *Elements* of Euclid, namely the equilateral triangle, the square, and the regular pentagon, hexagon, and 15-gon. Carl Friedrich Gauss (at age 19) proved that the regular 17-gon could be constructed with compass and straightedge, although he did not exhibit explicitly the construction. The regular 257-gon was constructed in 1832 by Friedrich Julius Richelot (1808–1875) and the regular 65537-gon was successfully constructed by Johann Gustav Hermes (1846–1912) in 1894 after a full ten years of dedication and hard work. This shows that sometimes patience and perseverance are welcome complements to mathematical ingenuity.

4.3   The theorems of Maekawa and Kawasaki

Origami (from the Japanese *oru*, "to fold," and *kami*, "paper") is a traditional Japanese art of folding a sheet of paper, usually square, into a representation of an object such as a bird or flower. *Flat origami* refers to configurations that can be pressed flat, say between the pages of a book, without adding any new folds or creases. When an origami object is unfolded, the resulting diagram of folds or creases on the paper square is called a *crease pattern*. In a crease pattern we see two types of folds, called *mountain folds* and *valley folds*. See Figure 4.6.

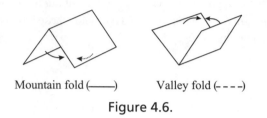

Mountain fold (——)   Valley fold (- - - -)

Figure 4.6.

We denote mountain folds by unbroken lines and valley folds by dashed lines. A *vertex* of a crease pattern is a point where two or more folds intersect, and a *flat vertex fold* is a crease pattern with just one vertex. In Figure 4.7 we see a classical origami figure—a crane—and a portion of its crease pattern.

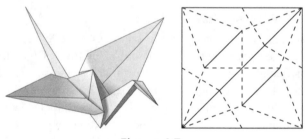

Figure 4.7.

We now consider two theorems about flat vertex folds, and their proofs. The first is named after the Japanese physicist Jun Maekawa. It first appeared in print with this name in 1987, but without proof. The French mathematician Jacques Justin published the result in 1986, also without proof. The following simple proof was found by Jan Siwanowicz while a high-school student [Hull, 1994].

Mackawa's Theorem 4.4. *The difference between the number of mountain folds and the number of valley folds in a flat vertex fold is two.*

Proof. Let n denote the number of folds that meet at the vertex, m of which are mountain folds and v that are valley folds, so that $n = m + v$. After folding the paper flat (see Figures 4.8a and 4.8b), cut the paper below the vertex to expose a folded polygonal cross-section, as in Figure 4.8c. This cross-section is a *flat polygon*, all of whose (interior) angles are 0° or 360°.

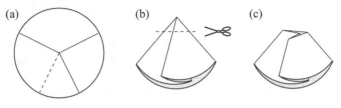

(a)   (b)   (c)

Figure 4.8.

Looking at the flat polygon from above, the mountain creases correspond to 0° angles and the valley creases to 360° angles, and hence the angle sum for the polygon is $(0m + 360v)°$. But Theorem 4.1(i) tells us that the angle sum is also $(n-2)180° = (m+v-2)180°$, so $m - v = 2$. If we had looked at the flat polygon from below (or had turned the paper over), we would have obtained $m - v = -2$. Hence $|m - v| = 2$ as claimed. ■

Since m and v differ by 2, they are either both even or both odd, and hence $n = m + v$ is even, which proves

Corollary 4.1. *The number of creases emanating from the vertex of a flat vertex fold is even.*

As a consequence the number of angles between successive creases is also even. The next theorem tells us about angle sums for flat vertex folds.

Kawasaki's Theorem 4.5. *The sum of the alternate angles about the vertex of a flat vertex fold is 180°.*

Proof. Let $\theta_1, \theta_2, \ldots, \theta_{2n}$ denote consecutive angles around the vertex. Of course we have $\theta_1 + \theta_2 + \cdots + \theta_{2n} = 360°$. If we fold the paper flat and trace a path on it around the vertex starting at one of the creases, each time we encounter a crease we reverse direction. After $2n$ steps we arrive back at our starting point, and thus we also have $\theta_1 - \theta_2 + \theta_3 - \cdots - \theta_{2n} = 0°$. Adding the two equations and dividing by 2 yields $\theta_1 + \theta_3 + \cdots + \theta_{2n-1} = 180°$, subtracting and dividing by 2 yields $\theta_2 + \theta_4 + \cdots + \theta_{2n} = 180°$. ■

The converse of Kawasaki's theorem is also true (see [Hull, 1994]), but these theorems do not tell the whole story. See Challenge 4.4. Furthermore, these results apply only to single-vertex flat folds; extending them to multi-vertex folds is difficult.

4.4   Squaring polygons

For the ancient Greek geometers, to *square* a plane figure meant constructing a square with the same area in a finite number of steps, using only a compass and straightedge. Over the centuries the problem of squaring figures bounded by curves (such as circles and lunes, see Chapter 9) has received much attention, but the fact that any polygon can be squared has been known since antiquity. Thus it is somewhat surprising that any regular polygon can be squared, but that the limiting figure, a circle, cannot be squared.

After a simple lemma for squaring triangles, we present two theorems that give two ways of squaring a polygon, the first direct and the second inductive.

Lemma 4.1. *Any triangle can be squared with straightedge and compass.*

Proof. Enclose the triangle in a rectangle, and then square the rectangle using the geometric construction for the geometric mean $\sqrt{ab}$ of two positive numbers a and b. See Figure 4.9. ∎

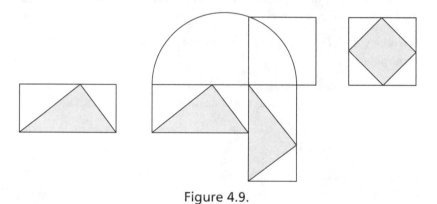

Figure 4.9.

Theorem 4.6. *Any convex n-gon can be squared with straightedge and compass.*

Proof. For $n = 3$ apply Lemma 4.1. For $n \geq 4$ partition the n-gon into $n - 2$ triangles by drawing diagonals from one vertex and apply Lemma 4.1 to each one to obtain $n - 2$ squares. Now sum the squares geometrically by repeatedly drawing right triangles and applying the Pythagorean theorem. ∎

Theorem 4.7. *Given a convex n-gon $P_n, n \geq 4$, there exists a convex $(n-1)$-gon P_{n-1} with the same area that can be constructed with straight-edge and compass.*

Proof. See Figure 4.10. ∎

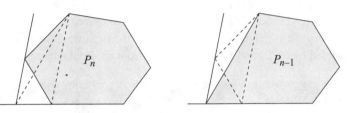

Figure 4.10.

In Chapter 9 we consider the problem of squaring figures bounded by curves.

4.5 The stars of the polygonal playground

Convex (and concave) polygons are *simple* in the sense that no one side of the polygon intersects another. When we allow intersections of the edges we obtain *star polygons*. Regular star polygons are denoted by the symbol $\{p/q\}$ where p and q are positive integers with $1 < q < p/2$. The regular star polygon $\{p/q\}$ is constructed by connecting every qth point out of p points evenly spaced on a circle (which is subsequently erased). In Figure 4.11 we have, from left to right, the star polygons $\{5/2\}, \{6/2\}, \{7/2\}, \{7/3\}, \{8/2\}$, and $\{8/3\}$.

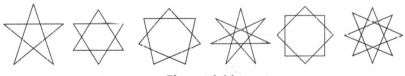

Figure 4.11.

Many star polygons have names: $\{5/2\}$ is a *pentagram*, $\{6/2\}$ the *hexagram, star of David*, or *Solomon's seal*, $\{8/2\}$ the *star of Lakshmi*, and $\{8/3\}$ the *octagram*. The pentagram has been used as a pagan religious symbol for over 5000 years and was also used as a religious symbol in early Christianity and in Islam. The pentagram appears on the national flags of Morocco and Ethiopia, the star of David on the national flag of Israel and on the flag of

British Nigeria prior to independence in 1960, and the octagram on the national flag of Azerbaijan and the national flag of the Republic of Iraq between 1959 and 1963. Star polygons appear frequently in Moorish ceramic tile patterns. On the left in Figure 4.12 we have an {8/2} star from the Patio del Cuarto Dorado in the Alhambra in Granada, Spain, built in the 14th century. In the tiling in the center in we see an {8/3} star within an {8/2} star, surrounded by eight more {8/3} stars, from the Real Alcázar in Seville, Spain, also built in the 14th century. In the tiling on the right we have a {12/4} star surrounded by {6/2} and {8/2} stars, also from the Real Alcázar.

Figure 4.12.

For general (i.e., not necessarily regular) star pentagons, we have the following theorem. Its proof is based on the one in [Nakhli, 1986].

Theorem 4.8. *The vertex angles of any star pentagon sum to* 180°.

Proof. See Figure 4.13.   ∎

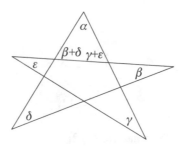

Figure 4.13.

The vertex angles of the star of David sum to 360° since the star is the union of two equilateral triangles. In general, the sum of the vertex angles of the regular star polygon {p/q} is (p − 2q)180°, a formula discovered by

Thomas Bradwardine (1290–1349), Archbishop of Canterbury [Eves, 1983].
See Challenge 4.6.

We now restrict our attention to polygons that are *ordinary* (no point in the
plane belongs to more than two edges) and *proper* (at each vertex adjacent
edges are non-collinear) [Grünbaum, 1975].

If we consider all possible pentagons (allowing for intersections of sides
and reflex angles), how many classes of pentagons are there? Following
[Steinitz, 1922], two ordinary and proper n-gons $P_0 = [a_1, a_2, \ldots, a_n]$ and
$P_1 = [b_1, b_2, \ldots, b_n]$ are *in the same class* if there exists a class of ordinary
and proper n-gons $P(t) = [x_1(t), x_2(t), \ldots, x_n(t)]$ for $0 \leq t \leq 1$ such that
$P(0) = P_0$, $P(1)$ coincides with P_1 or with a mirror image of P_1, and for
each i in $\{1, 2, \ldots, n\}$ the function $x_i(t)$ is continuous on [0,1]. Then there
are exactly 11 classes of pentagons [Grünbaum, 1975], and they are shown
in Figure 4.14.

Figure 4.14.

The complete classification of hexagons is unknown, but 72 classes have
been identified [Grünbaum, 1975].

Star polygons and star-shaped polygons

A *star-shaped region* (or *star domain* or *star-convex set*) in the plane is
a set S of points with the property that there exists a point A in S such
that for any point X in S, the line segment AX lies entirely within S.
When S is a star-shaped polygonal region, we call it a *star-shaped poly-
gon*. While every star polygon is a star-shaped polygon, there are star-
shaped polygons that are not star polygons (e.g., convex polygons). The
class of star-shaped polygons is used in the study of visibility problems
(see the next section) and robotics.

4.6   Guards in art galleries

In 1973 Victor Klee formulated the celebrated art gallery problem [Aigner and Zeigler, 2001]: How many guards do we need in an art gallery to be certain that they, looking in all directions, can watch all points of the gallery?

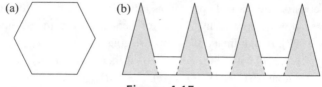

Figure 4.15.

We consider the case of a polygonal art gallery. When the polygon is convex, clearly one guard suffices (see Figure 4.15a). However, for concave polygons more that one guard may be needed. In Figure 4.15b we see a comb-shaped gallery with twelve walls that requires four guards, one located in each shaded triangle in order to watch each tooth of the comb. In general, we have the following theorem, first proved in [Chvátal, 1975]. Our proof [Do, 2004] is based on the elegant one-page proof in [Fisk, 1978].

Theorem 4.9. *For any polygonal art gallery with n walls, $\lfloor n/3 \rfloor$ guards always suffice and may be necessary.*

Proof. The necessity follows from comb-shaped galleries with k teeth, similar to the one in Figure 4.15b, that have $3k$ walls and require k guards. By clipping off a corner or two, we have galleries with $3k + 1$ or $3k + 2$ walls that require k guards.

The proof that $\lfloor n/3 \rfloor$ guards always suffice has three main steps.

(i) *We can always triangulate a polygon.* Triangulation of a polygon means partitioning it into non-overlapping triangles by means of nonintersecting diagonals that join pairs of vertices. This can be done if every polygon always has at least one diagonal joining two vertices that lies entirely within the polygon. If this is true, then we use it to cut the polygon into two smaller polygons, and repeat the process until we are left with only triangles.

To see that there always exists such a diagonal (for $n \geq 4$), we station a guard at vertex A with a flashlight shining towards an adjacent vertex B. If the guard now rotates the flashlight towards the interior of the polygon, then it must eventually shine on another vertex C, and we can take AC or BC as the desired diagonal, as long as there is no other vertex of the polygon inside triangle ABC. But no such vertex exists, for otherwise the flashlight would have shined upon it before reaching C.

(ii) *Every triangulation of a polygon can be 3-colored.* A 3-coloring of a polygon triangulation is a way to assign one of three colors to each vertex of the polygon so that each triangle in the triangulation has vertices of three different colors. Since it is trivial to 3-color a triangle, we consider an n-gon with $n \geq 4$. With the result above, we can find a diagonal joining two vertices that lies entirely within the polygon, which will partition the polygon into two smaller ones. If we can 3-color the two smaller polygons, we can join them together, after perhaps relabeling the colors in one of the pieces (see Figure 4.16). We can 3-color the smaller polygons by repeating the process, partitioning them into smaller and smaller polygons until we have only triangles.

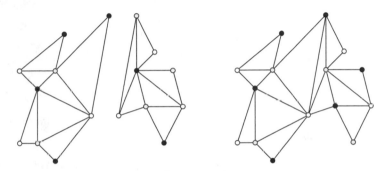

Figure 4.16.

(iii) *Now place guards at the vertices with the minority color.* Suppose the colors of the vertices are red, blue, and yellow. If we place guards at all the vertices of one color, then clearly the entire gallery is guarded. So we chose the color that appears least often. Since it is impossible for each of the three colors to appear more than $n/3$ times, one color (say red) appears in the 3-coloring k times, $k \leq n/3$. But since k is an integer, we have $k \leq \lfloor n/3 \rfloor$. ■

4.7 Triangulation of convex polygons

The problem of counting all possible triangulations of a given convex n-gon has a distinguished history. The problem (and tentative answers for several values of n) was stated by Leonhard Euler (1707–1783) in a letter to Christian Goldbach (1690–1764) [Euler, 1965]. Euler also communicated the problem to Jan Andrej Segner (1704–1777), who solved it. Later it was presented as an open challenge by Joseph Liouville (1809–1894) and solved by several mathematicians, including Gabriel Lamé (1795–1870). In this section we present Lamé's combinatorial arguments [Lamé, 1838]. Triangulating a polygon means partitioning it into non-overlapping triangles by means of nonintersecting diagonals that join pairs of vertices of the polygon. For example, the number of triangulations of a square, a pentagon, and a hexagon are 2, 5, and 14, respectively. See Figure 4.17.

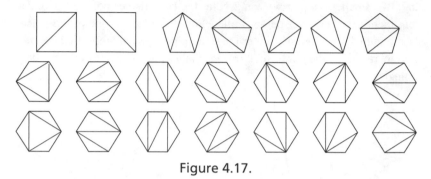

Figure 4.17.

Theorem 4.10. *Let $T_n, n \geq 3$, denote the number of triangulations of a convex n-gon. Then $T_3 = 1$ and*

(i) $T_{n+1} = T_n + T_3 T_{n-1} + T_4 T_{n-2} + \cdots + T_{n-2} T_4 + T_{n-1} T_3 + T_n$ *for $n \geq 3$,*

(ii) $T_n = n (T_3 T_{n-1} + T_4 T_{n-2} + \cdots + T_{n-2} T_4 + T_{n-1} T_3)/(2n - 6)$ *for $n \geq 4$, and*

(iii) $T_{n+1} = (4n - 6) T_n / n$ *for $n \geq 3$.*

Proof. (i) We illustrate the proof with an octagon, but it is clear how the proof proceeds for an n-gon. See Figure 4.18.

Figure 4.18.

Consider the bottom side of the octagon. After triangulation, it is the base of exactly six triangles, as illustrated in the figure. In each case there is a polygon (possibly just a line segment or a "2-gon") to the left and one to the right of the triangle that must also be triangulated.

In the first octagon in Figure 4.18, there is a line segment on the left and a 7-gon the right, which can be triangulated in T_7 ways, In the second octagon, there is a triangle on the left and a hexagon on the right, which together can be triangulated in $T_3 T_6$ ways. Continuing in this fashion for the remaining four figures yields $T_8 = T_7 + T_3 T_6 + T_4 T_5 + T_5 T_4 + T_6 T_3 + T_7$.

(ii) We illustrate the proof with a heptagon, but it is clear how the proof proceeds for an n-gon. See Figure 4.19. Each diagonal appears in many triangulations. The leftmost diagonal in Figure 4.19 appears in $T_3 T_6$ triangulations, as it partitions the heptagon into a triangle and a hexagon. The diagonal to its right appears in $T_4 T_5$ triangulations, as it partitions the heptagon into a quadrilateral and a pentagon. Since the heptagon has seven vertices, the quantity $L_7 = 7 (T_3 T_6 + T_4 T_5 + T_5 T_4 + T_6 T_3)$ includes all possible triangulations counted by means of the diagonals, but of course many of them have been counted several times. In general, we have $L_n = n(T_3 T_{n-1} + T_4 T_{n-2} + \cdots + T_{n-1} T_3)$.

Figure 4.19.

Each triangulation of an n-gon has $n - 3$ different diagonals, and any given set of $n - 3$ different diagonals will appear in the sum L_n exactly $2(n - 3)$ times, since each diagonal has two endpoints. Thus $T_n = L_n/(2n - 6)$ as claimed. Finally, (iii) is an immediate consequence of (i) and (ii). ■

The result in part (iii) of Theorem 4.10 yields an explicit formula for T_n as given in Corollary 4.2. We omit the proof, which is by mathematical induction.

Corollary 4.2. *Let T_n, $n \geq 3$, denote the number of triangulations of a convex n-gon. Then*

$$T_n = \frac{1}{n - 1} \binom{2n - 4}{n - 2}.$$

Catalan numbers

The numbers $\{T_n\}_{n=3}^{\infty}$ are better known as the *Catalan numbers* $C_n = T_{n+2}$, i.e., the nth Catalan number is the number of ways to triangulate a $(n + 2)$-gon for $n \geq 1$. They also appear in the solutions to many other combinatorial problems, such as parentheses in products, diagonal-avoiding paths in a square lattice, and binary trees. They are named after Eugène Charles Catalan (1814–1894), a French mathematician born in Bruges (when it was part of France). He worked in number theory (particularly continued fractions), descriptive geometry, and combinatorics. He is also responsible for *Catalan's conjecture* that 8 and 9 are the only nontrivial consecutive integer powers (the first proof of which was published in 2004 by Preda Mihăilescu).

4.8   Cycloids, cyclogons, and polygonal cycloids

It is well known that when one rolls a circle along a line, the curve generated by a point on the circle is the *cycloid*. Two of the nice properties of a cycloid are: (i) the area under a cycloid is three times the area of the generating circle; and (ii) the length of a cycloid is four times the diameter of the generating circle. It may be surprising that similar results hold if the circle is replaced by a regular polygon.

If we replace the circle by a regular polygon, the curve generated by a vertex of the polygon consists of arcs of circles. Such a curve is sometimes called a *cyclogon* [Apostol and Mnatsakanian, 1999]. See Figure 4.20a for the case of a hexagon rolling along a line.

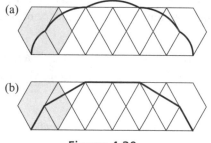

Figure 4.20.

If we replace the arcs by their chords, the resulting figure is called a *polygonal cycloid*, and is illustrated for the hexagon in Figure 4.20b.

We now find the area under one arch of a polygonal cycloid, and its length. But first we prove a lemma, which is of interest in its own right.

Lemma 4.2. *If $V_1, V_2, \ldots, V_n$ are the vertices of a regular n-gon with circumradius R, and if P is any point on the circumcircle of the n-gon, then*

$$|PV_1|^2 + |PV_2|^2 + \cdots + |PV_n|^2 = 2nR^2.$$

Proof [Ouellette and Bennett, 1979]. Place the n-gon in the xy-plane so that the center of the circumcircle is at the origin and let $V_i = (a_i, b_i)$ and $P = (u, v)$. Then

$$
\begin{aligned}
|PV_1|^2 + |PV_2|^2 + \cdots + |PV_n|^2 &= \sum_1^n (u - a_i)^2 + \sum_1^n (v - b_i)^2 \\
&= n(u^2 + v^2) - 2u \sum_1^n a_i - 2v \sum_1^n b_i \\
&\quad + \sum_1^n (a_i^2 + b_i^2) \\
&= 2nR^2 - 2u \sum_1^n a_i - 2v \sum_1^n b_i,
\end{aligned}
$$

since $u^2 + v^2 = R^2$ and $a_i^2 + b_i^2 = R^2$. To complete the proof we need only show that $\sum_1^n a_i = \sum_1^n b_i = 0$. Place equal weights at each of the vertices of the n-gon. Since the center of gravity of the n weights will be the center of the circumcircle, the x and y moments of the system are each zero, hence $\sum_1^n a_i = \sum_1^n b_i = 0$. (A formal proof using complex numbers can be found in [Ouellette and Bennett, 1979]). ■

A special case of interest is when the point P is one of the vertices of the n-gon. Then we have

Corollary 4.3. *The sum of the squared distances from one vertex of a regular n-gon with circumradius R to each of the other $n - 1$ vertices is $2nR^2$.*

We now prove that the area property for the cycloid mentioned above also holds for polygonal cycloids.

Theorem 4.11. *When a regular polygon is rolled along a line, the area of the polygonal cycloid generated by a vertex of the polygon is three times the area of the polygon.*

Proof. Let R denote the circumradius of the n-gon, and let $d_1, d_2, \ldots, d_{n-1}$ denote the distances from a given vertex (say V_1) of the n-gon to the other $n - 1$ vertices, as illustrated in Figure 4.21a.

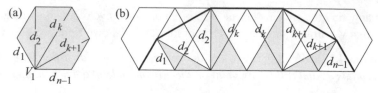

Figure 4.21.

Let A be the numerical value of the area of the regular n-gon with circumradius 1. Figure 4.21b shows the polygonal cycloid generated by V_1 as the polygon rolls along the line. The region under the polygonal cycloid can be partitioned into $n - 2$ shaded triangles and $n - 1$ white isosceles triangles, each with apex angle $2\pi/n$. The equal-length sides in the isosceles triangles are successively $d_1, d_2, \ldots, d_{n-1}$. The $n - 2$ shaded triangles are congruent to the $n - 2$ triangles formed by the diagonals in the original n-gon, and hence their areas sum to $R^2 A$. Using Corollary 4.2, the sum of the areas of the $n - 1$ white isosceles triangles is

$$\frac{1}{n} A \left(d_1^2 + d_2^2 + \cdots + d_{n-1}^2 \right) = \frac{A}{n} \cdot 2nR^2 = 2R^2 A.$$

Hence the area under the polygonal cycloid is $3R^2 A$, or three times the area of the generating n-gon. ∎

We now turn our attention to the length of a polygonal cycloid.

Theorem 4.12. *When a regular polygon is rolled along a line, the length of the polygonal cycloid generated by a vertex of the polygon is four times the sum of the inradius and the circumradius of the polygon.*

Proof [Mallinson, 1998b]. We treat the cases n even and n odd separately. In each case we draw rhombi whose side lengths are the same as the side length of the n-gon, and whose diagonals are segments of the polygonal cycloid. These rhombi can then be arranged in the n-gon to show that the sum of the lengths of the segments of the polygonal cycloid is four times the sum of the inradius and the circumradius of the generating n-gon. In Figure 4.22 we illustrate the $n = 10$ case; and in Figure 4.23 the $n = 9$ case. ∎

The same approach—using rhombi and their diagonals—can be used to prove Theorem 4.11 [Mallinson, 1998a].

At the beginning of this section we noted that the curve generated by the vertex of a polygon as it rolls along a line is a cyclogon, a collection

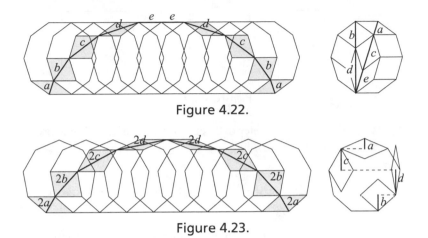

Figure 4.22.

Figure 4.23.

of circular arcs. The area under a cyclogon is easily computed using methods similar to those found in the proof of Theorem 4.11. See [Apostol and Mnatsakanian, 1999] for a proof.

Theorem 4.13. *When a regular polygon is rolled along a line, the area of the cyclogon generated by a vertex of the polygon is equal to the area of the polygon plus twice the area of the circumscribed circle of the polygon.*

4.9 Challenges

4.1 Let $n \geq 3$, If p_n and P_n (a_n and A_n) denote respectively the perimeters (areas) of regular n-gons inscribed in and circumscribed about the same circle, show that $p_n/P_n = \cos \pi/n$ and $a_n/A_n = \cos^2 \pi/n$.

4.2 Prove that the area a_{2n} of the regular $2n$-gon inscribed in the unit circle is numerically the same as half the perimeter p_n of the inscribed regular n-gon.

4.3 If p_k and P_k denote respectively the perimeters of regular k-gons inscribed in and circumscribed about the same circle, show that

$$P_{2n} = 2p_n P_n/(p_n + P_n) \quad \text{and} \quad p_{2n} = \sqrt{p_n P_{2n}}.$$

4.4 In flat origami the Maekawa and Kawasaki theorems tell us that the number of mountain and valley folds and the angles between them are critical. But so is the order in which the fold appear around the vertex. The crease pattern in Figure 4.24a folds flat, as illustrated in Figure 4.24b, but the pattern in Figure 4.24c does not. Why not?

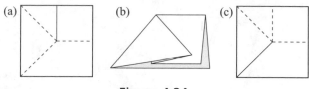

Figure 4.24.

4.5 Let p_n denote the perimeter of a regular n-gon inscribed in a circle of radius r. Use the fact that $\lim_{n\to\infty} p_n = 2\pi r$ to prove $\lim_{\theta\to 0+} \sin\theta/\theta = 1$.

4.6 Prove that the sum of the vertex angles of the regular star polygon $\{p/q\}$ is $(p-2q)180°$.

4.7 Let P be a regular n-gon with circumradius R. Show that the sum of the squares of the lengths of all the sides and all the diagonals of P is $n^2 R^2$.

4.8 In a given convex n-gon let T_n denote the number of triangulations of the n-gon. Show that $T_n \geq 2T_{n-1}$ and $T_n \geq 2^{n-3}$.

4.9 What are the measures of the angles in a polygonal cycloid generated by rolling an n-gon on a line?

4.10 Let Q be a point on or in a regular n-gon. Prove that the sum of the perpendicular distances from Q to the sides of the n-gon is constant. [Hint: the constant is n times the inradius of the n-gon.]

4.11 (a) Show that a regular hexagon, six squares, and six equilateral triangles can be assembled without overlapping to form a regular dodecagon.

(b) Let $V_1, V_2, \ldots, V_{12}$ be the successive vertices of a regular dodecagon. Discuss the intersection(s) of the three diagonals $V_1 V_9$, $V_2 V_{11}$, and $V_4 V_{12}$ (Problem I-1, 24th Annual William Lowell Putnam Mathematical Competition, 1963).

CHAPTER 5

A Treasury of Triangle Theorems

The only royal road to elementary geometry is ingenuity.
Eric Temple Bell

The present author humbly confesses that, to him, geometry is nothing at all, if not a branch of art.
Julian L. Coolidge

Book I of the *Elements* of Euclid is devoted to theorems about parallel lines, area, and triangles. Of the 48 propositions in this book, 23 concern the triangle. Proposition 47 in Book I is perhaps the best-known theorem in mathematics, the Pythagorean theorem. Consequently one can justifiably say that triangles lie at the very core of geometry.

We begin this chapter with several proofs of the Pythagorean theorem, followed by some related results, including Pappus' generalization of the Pythagorean theorem. Consideration of the inscribed and circumscribed circles for general triangles leads to Heron's formula and Euler's inequality. We conclude this chapter with the Erdős-Mordell inequality, the Steiner-Lehmus theorem, some results about triangle medians, and a Lewis Carroll problem.

5.1 The Pythagorean theorem

As we mentioned in the Introduction, there may well be more proofs of the Pythagorean theorem than of any other theorem in mathematics. The classic book *The Pythagorean Proposition* by Elisha Scott Loomis [Loomis, 1968] presents 370 proofs. Alexander Bogomolny's website *Interactive Mathematics Miscellany and Puzzles*, www.cut-the-knot.org, has 84 proofs (as of

2009), many of them interactive. An argument can be made that the best proof is Euclid's, where the theorem appears as Proposition 47 in Book I of the *Elements* [Joyce, 1996]: *In right-angled triangles the square on the side opposite the right angle equals the sum of the squares on the sides containing the right angle.*

The double-counting combinatorial proof technique mentioned in the Introduction can be extended to areas: *computing the area of a region in two different ways yields the same number.* This extension of the Fubini principle is used in a classic proof of the Pythagorean theorem. This proof is based on the one from the *Zhou bi suan jing*, a Chinese document dating from approximately 200 BCE. See Figure 5.1.

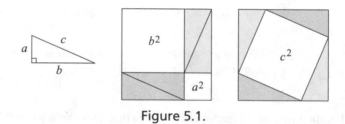

Figure 5.1.

For a right triangle with legs a and b and hypotenuse c, we express the area of a square with side length $a + b$ in two different ways to obtain $a^2 + b^2 = c^2$.

Another way to use the extended Fubini principle to prove the Pythagorean theorem is by *dissection*: cut the squares with areas a^2 and b^2 into several pieces, and reassemble those pieces to form a square with area c^2. The dissection proof in Figure 5.2a is usually attributed to Annairizi of Arabia (c. 900 AD), while the one in Figure 5.2b is often attributed to Henri Perigal (1801–1899).

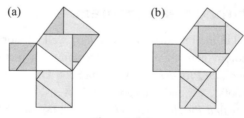

Figure 5.2.

5.2 Pythagorean relatives

In this section we present a variety of results related to the Pythagorean theorem.

The Reciprocal Pythagorean Theorem 5.1. *If a and b are the legs and h the altitude to the hypotenuse of a right triangle, then*

$$\left(\frac{1}{a}\right)^2 + \left(\frac{1}{b}\right)^2 = \left(\frac{1}{h}\right)^2.$$

Proof. Let c be the length of the hypotenuse. Multiplying the sides by $1/ab$ yields a triangle with sides $1/b, 1/a, c/ab$ similar to the triangle with sides a, b, c. Computing the area of the given triangle in two ways yields $ab/2 = ch/2$, hence $c/ab = 1/h$. Thus the triangle with sides $1/b, 1/a, 1/h$ is a right triangle (and similar to the given triangle).   ■

A Pythagorean-like Theorem 5.2 [Hoehn, 2000]. *In an isosceles triangle, let c denote the length of the equal sides and draw a line segment of length a from the apex of the triangle to the third side dividing that side into segments of lengths b and d, as shown in Figure 5.3a. Then $c^2 = a^2 + bd$.*

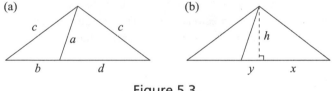

Figure 5.3.

Proof. Let h denote the altitude of the triangle, and set $d = x + y$ and $b = x - y$ as shown in Figure 5.3b. Two applications of the Pythagorean theorem yield $c^2 = h^2 + x^2$ and $a^2 = h^2 + y^2$, hence $c^2 = a^2 + x^2 - y^2$. But $x^2 - y^2 = (x - y)(x + y) = bd$, which completes the proof.   ■

The Right Angle Bisector Theorem 5.3 [Eddy, 1991]. *The internal bisector of the right angle of a right triangle bisects the square on the hypotenuse* (See Figure 5.4a).

Proof. See Figure 5.4b, which is derived from the *Zhou bi suan jing* proof of the Pythagorean theorem (Figure 5.1). The figure also proves that the

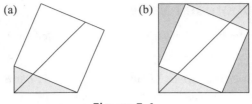

Figure 5.4.

bisector of the right angle passes through the center of the square on the hypotenuse. ∎

It is easy to show that the only right triangles with sides in an arithmetic progression are the triangles similar to the 3:4:5 right triangle. What about sides in a geometric progression? The following theorem provides the somewhat surprising answer [Herz-Fischler, 1993]. Its proof follows directly from the Pythagorean theorem.

Theorem 5.4. *The sides of a right triangle are in geometric progression if and only if the triangle is similar to the triangle with sides* 1, $\sqrt{\varphi}$, *and* φ, *where* φ *is the golden ratio.*

Johannes Kepler (1571–1630) wrote "Geometry has two great treasures: one is the theorem of Pythagoras; the other the division of a line into extreme and mean [i.e., golden] ratio. The first we may compare to a measure of gold, the second we may name a precious jewel." The triangles in Theorem 5.4 are sometime called *Kepler triangles*, as they combine the two treasures in the Kepler quotation.

Kepler triangles and the Great Pyramid

The Great Pyramid of Khufu in Giza, Egypt is home to much speculation and some surprising coincidences. Some pyramid enthusiasts claim that the Great Pyramid was built so that the area of each triangular face is equal to the square of the height of the pyramid (see Figure 5.5). If this were true, then letting b denote the side of the square base, h the height, and s the altitude of a triangular face yields $bs/2 = h^2$ and $s^2 = h^2 + (b/2)^2$. For convenience set $b = 2$ so that $s = h^2$ and $s^2 = h^2 + 1$, thus $s^2 = s + 1$. Hence s is the golden ratio φ, and the dark gray right triangle in Figure 5.5 with sides s, $b/2$, and h is a Kepler triangle.

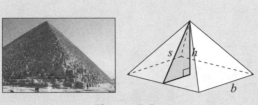

Figure 5.5.

Other pyramid mystics claim that the pyramid was constructed so that the perimeter of the base equals the circumference of a circle whose radius is the height, or $4b = 2\pi h$. If both claims are true (which, of course, they are not), then with $b = 2$ and $h = \sqrt{\varphi}$ we have $\pi = 4/\sqrt{\varphi} \approx 3.1446$, which is accurate to within 0.1%, a remarkable coincidence [Peters, 1978]. For another relationship between π and φ, see Section 12.3.

5.3 The inradius of a right triangle

It is easy to compute the length r of the *inradius* (the radius of the inscribed circle) of a right triangle. As before, we let a and b denote the legs, c the hypotenuse, and r the inradius of a right triangle. See Figure 5.6a. Then r can be easily computed by one of

$$r = \frac{a + b - c}{2} \quad \text{and} \quad r = \frac{ab}{a + b + c}. \tag{5.1}$$

In Figure 5.4b, we see that $c = a + b - 2r$, and hence $r = (a + b - c)/2$. Using the decomposition of the triangle illustrated in Figure 5.5b, we see that pieces of a rectangle with area ab can be rearranged to form a rectangle with area $r(a + b + c)$ in Figure 5.6c, from which we obtain $r = ab/(a + b + c)$.

Finally, if we eliminate r from the two equations in (5.1), we obtain $a^2 + b^2 = c^2$—yet another proof of the Pythagorean theorem! We also obtain another formula for the area K of the right triangle. Since $K = ab/2$ and $ab = r(a + b + c)$, we have $K = rs$ where s denotes the *semiperimeter* $(a + b + c)/2$. This result actually holds for any triangle—see Lemma 5.1 in Section 5.5.

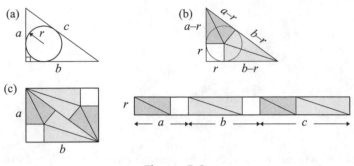

Figure 5.6.

The next theorem presents a curious area formula for right triangles.

Theorem 5.5. *The area K of a right triangle is equal to the product xy of the lengths of the segments of the hypotenuse determined by the point of tangency of the inscribed circle* (see Figure 5.7a).

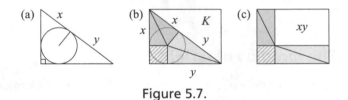

Figure 5.7.

Proof. Since $x = a - r$ and $y = b - r$ (see Figure 5.6b), an algebraic proof exists (see Challenge 5.2) but the following visual proof is nicer. First construct a rectangle from two copies of the triangle (see Figure 5.7b). Then rearrange the portions of the shaded triangle as shown in Figure 5.7c. Since the areas of the large rectangle and the shaded areas are the same in both figures, so are the areas of the unshaded portions. Thus $K = xy$. ∎

We conclude this section with an intriguing result concerning three circles inscribed in a right triangle.

Theorem 5.6. *Let ABC be a right triangle with inradius r, let h be the length of the altitude CD to the hypotenuse, and let r′ and r″ denote the inradii of triangles ACD and BCD* (see Figure 5.8). *Then*

$$r + r' + r'' = h.$$

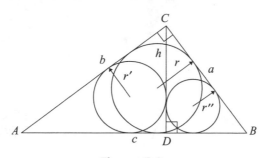

Figure 5.8.

Proof. Since $|AD| = b^2/c$ and $|BD| = a^2/c$, we can apply the first equality in (5.1) to each of the three right triangles to obtain

$$r + r' + r'' = (1/2)\left[(a + b - c) + (b^2/c + h - b) + (a^2/c + h - a)\right]$$
$$= (1/2)\left[(a^2 + b^2)/c - c + 2h\right] = h. \qquad \blacksquare$$

5.4 Pappus' generalization of the Pythagorean theorem

In Book IV of his *Mathematical Collection*, Pappus of Alexandria (c. 320 AD) gives the following generalization of the Pythagorean theorem. It generalizes the theorem in two directions—the triangle need not be a right triangle, and one constructs parallelograms on the sides, rather than squares.

Pappus' Area Theorem 5.7. *Let ABC be any triangle, and $ABDE$ and $ACFG$ any parallelograms constructed externally on AB and AC, respectively* (see Figure 5.9). *Extend DE and FG to meet in H, and draw BL and CM equal and parallel to HA. Then, in area, $BCML = ABDE + ACFG$.*

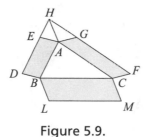

Figure 5.9.

Proof. Our proof of this theorem is dynamic, which we can only describe here. It is adapted from the simple proof in [Eves, 1969]. See Figure 5.10.

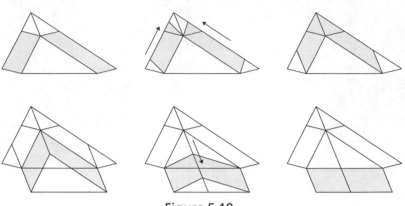

Figure 5.10.

We employ shear transformations on parallelograms $ABDE$ and $ACFG$ so that vertices E and G coincide with H; in the process the areas remain the same. The two resulting parallelograms are then translated downwards a distance $HA = BL = CM$; then sheared again to form parallelogram $BCML$. Once again the area remains the same, completing the proof.   ∎

5.5 The incircle and Heron's formula

Heron of Alexandria's (c. 10–75 AD) remarkable formula for the area K of a triangle with sides a, b, and c is $K = \sqrt{s(s-a)(s-b)(s-c)}$ where s denotes the *semiperimeter* $(a+b+c)/2$. It is perhaps the easiest way to find the area of a triangle given the lengths of the three sides. For example, if $(a,b,c) = (13, 14, 15)$, then $K = \sqrt{21 \cdot 8 \cdot 7 \cdot 6} = 84$.

We begin by bisecting each angle of the triangle to locate the incenter I, and then use the common perpendicular distance to each side as the inradius, as shown in Figure 5.11a. The three angle bisectors and three inradii partition the triangle into three pairs of congruent right triangles, as shown

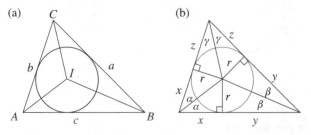

Figure 5.11.

in Figure 5.11b. In the figure we let $a = y + z$, $b = x + z$, and $c = x + y$. Then $s = x + y + z$, $x = s - a$, $y = s - b$, and $z = s - c$. We also let $A = 2\alpha$, $B = 2\beta$, and $C = 2\gamma$.

We now prove two lemmas. The first expresses the area K in terms of r and s while the second, whose proof uses a rectangle composed of triangles similar to the right triangles in Figure 5.11b, expresses the product xyz in terms of r and s.

Lemma 5.1. $K = r(x + y + z) = rs$.

Proof. See Figure 5.12. ∎

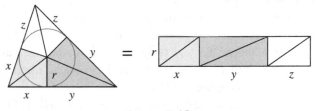

Figure 5.12.

Lemma 5.2. $xyz = r^2(x + y + z) = r^2s$.

Proof. See Figure 5.13, where w denotes $\sqrt{r^2 + x^2}$. To obtain the desired result, equate the heights on the left and right sides of the rectangle. ∎

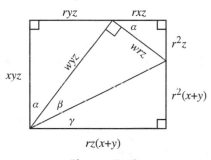

Figure 5.13.

We now have

Theorem 5.8 (Heron's formula). *The area K of a triangle with sides a, b, and c is given by*

$$K = \sqrt{s(s - a)(s - b)(s - c)}$$

where $s = (a + b + c)/2$.

Proof. From Lemma 5.2, we have $sxyz = r^2s^2 = (rs)^2$, but $K = rs$ from Lemma 5.1, hence $K^2 = sxyz = s(s-a)(s-b)(s-c)$, as desired. ∎

5.6 The circumcircle and Euler's triangle inequality

As with the incircle, every triangle also possesses a circumscribed circle (the *circumcircle*, whose center is the *circumcenter* of the triangle and whose radius is the *circumradius* of the triangle), exterior to the triangle and passing through each of the three vertices. If we let R denote the circumradius, then Euler's celebrated triangle inequality states that for any triangle, $R \geq 2r$.

It may be surprising that the inequality in Euler's triangle inequality derives from the *arithmetic mean-geometric mean inequality* (or *AM-GM inequality*) for two numbers: if x and y are nonnegative, then their arithmetic mean $(x + y)/2$ is greater than or equal to their geometric mean $\sqrt{xy}$. We extend this to n numbers in Section 12.4.

Applying the AM-GM inequality to each side of the triangle in Figure 5.12 yields the inequality $abc \geq 8xyz$:

$$abc = (y+z)(x+z)(x+y) \geq 2\sqrt{yz} \cdot 2\sqrt{xz} \cdot 2\sqrt{xy} = 8xyz. \quad (5.3)$$

In Lemma 5.2 we established a relationship between the product xyz and the inradius r. In the following lemma we establish a similar relationship between the product abc and circumradius R.

Lemma 5.3. $abc = 4KR$.

Proof. See Figure 5.14. By similar triangles (shaded gray), $h/b = a/2R$ and hence $h = ab/2R$. Thus $K = hc/2 = abc/4R$. ∎

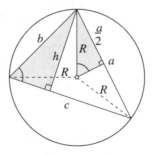

Figure 5.14.

We now have

Theorem 5.9 (Euler's triangle inequality). *If r and R denote the inradius and circumradius, respectively, of a triangle, then $R \geq 2r$.*

Proof. Using (5.3) and Lemmas 5.1, 5.2, and 5.3, we have

$$\frac{R}{r} = \frac{4KR}{4Kr} = \frac{abc}{4r^2 s} \geq \frac{8xyz}{4xyz} = 2. \qquad \blacksquare$$

We can express the inequality (5.3) solely in terms of a, b, and c by noting that $2x = b + c - a$, $2y = a + c - b$, and $2z = a + b - c$. Doing so yields

$$abc \geq (a + b - c)(a + c - b)(b + c - a).$$

This is *Padoa's inequality* (Alessandro Padoa, 1868–1937).

5.7 The orthic triangle

In 1775, Giovanni Francesco Fagnano dei Toschi (1715–1797) posed the following problem: given an acute triangle, find the inscribed triangle of minimum perimeter. By *inscribed triangle* in a given triangle ABC, we mean a triangle PQR such that each vertex P, Q, R lies on a different side of ABC. Fagnano solved the problem using calculus, but we present a clever non-calculus solution using reflection and symmetry due to Lipót Fejér (1880–1959) [Kazarinoff, 1961]. The solution involves the *orthic triangle*—the triangle PQR whose vertices are the feet of the altitudes from each of the vertices of ABC, as illustrated in Figure 5.15a.

Theorem 5.10. *In any acute triangle, the inscribed triangle with minimum perimeter is the orthic triangle.*

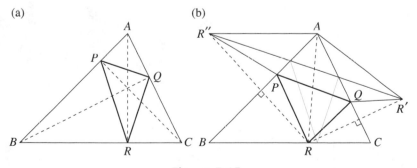

Figure 5.15.

Proof. To find the locations for points P, Q, R that minimize the perimeter of PQR, reflect point R in sides AB and AC (as in the previous section) to locate points R' and R''. Thus the perimeter of PQR is equal to $|R''P| + |PQ| + |QR'|$. The perimeter of PQR will be a minimum whenever R'', P, Q, and R' all lie on the same line. So for any given R, this gives the optimal location for P and Q. To find the optimal location for R, note that triangle $R''AR'$ is isosceles, with $|AR''| = |AR'| = |AR|$ and vertex angle $\angle R''AR' = 2\angle BAC$. Since the size of the vertex angle of triangle $R''AR'$ does not depend on R, the base $R''R'$ (the perimeter of PQR) will be shortest when the legs are shortest, and the legs are shortest when $|AR|$ is a minimum, i.e., when AR is perpendicular to BC. ∎

5.8   The Erdős-Mordell inequality

In 1935, the following problem proposal appeared in the Advanced Problems section of the *American Mathematical Monthly* [Erdős, 1935]:

3740. *Proposed by Paul Erdős, The University, Manchester, England.*

From a point O inside a given triangle ABC the perpendiculars OP, OQ, OR are drawn to its sides. Prove that

$$|OA| + |OB| + |OC| \geq 2(|OP| + |OQ| + |OR|).$$

Trigonometric solutions by Mordell and Barrow appeared in [Mordell and Barrow, 1937]; the proofs, however, were not elementary. Since then a variety of proofs have appeared, each in some sense simpler or more elementary than the preceding ones. The following simple visual proof appeared in [Alsina and Nelsen, 2007].

In Figure 5.16a we see the triangle as described by Erdős, and in Figure 5.16b we denote the lengths of relevant line segments by lower case

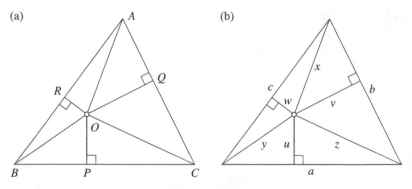

Figure 5.16.

letters. In terms of that notation, the Erdős-Mordell inequality becomes $x + y + z \geq 2(u + v + w)$.

Using similar figures, we construct a trapezoid in Figure 5.17b from three triangles—one similar to ABC, the other two similar to the shaded triangles in Figure 5.17a—to prove Lemma 5.4. Our proof applies to acute triangles; when the triangle is obtuse, an analogous construction using similar triangles yields the same inequalities.

Lemma 5.4. *For the triangle ABC in Figure* 5.16b, *we have* $ax \geq bw + cv$, $by \geq aw + cu$, *and* $cz \geq av + bu$.

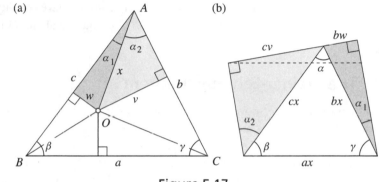

Figure 5.17.

Proof. The dashed line segment in Figure 5.17b has length ax, and hence $ax \geq bw + cv$. The other two inequalities are established analogously. ∎

We should note that the object in Figure 5.17b really is a trapezoid: the three angles at the point where the three triangles meet measure $(\pi/2) - \alpha_2, \alpha = \alpha_1 + \alpha_2$, and $(\pi/2) - \alpha_1$, and thus sum to π.

We now prove

The Erdős-Mordell Theorem 5.11. *If O is a point within a triangle ABC whose distances to the sides are* u, v, *and* w *and whose distances to the vertices are* x, y, *and* z, *then*

$$x + y + z \geq 2(u + v + w).$$

Proof. From Lemma 5.4 we have $x \geq \dfrac{b}{a}w + \dfrac{c}{a}v$, $y \geq \dfrac{a}{b}w + \dfrac{c}{b}u$, and $z \geq \dfrac{a}{c}v + \dfrac{b}{c}u$; adding these yields

$$x + y + z \geq \left(\frac{b}{c} + \frac{c}{b}\right)u + \left(\frac{u}{c} + \frac{c}{a}\right)v + \left(\frac{a}{b} + \frac{b}{a}\right)w. \tag{5.4}$$

But the AM-GM inequality insures that the coefficients of u, v, and w are each at least 2, from which the desired result follows. ∎

The three inequalities in Lemma 5.4 are equalities if and only if O is the center of the circumscribed circle of ABC. This follows from the observation that the trapezoid in Figure 5.17b is a rectangle if and only if $\beta + \alpha_2 = \pi/2$ and $\gamma + \alpha_1 = \pi/2$ (and similarly in the other two cases), so $\angle AOQ = b = \angle COQ$. Hence the right triangles AOQ and COQ are congruent, so $x = z$. Similarly one can show that $x = y$, hence $x = y = z$ so O must be the circumcenter of ABC. The coefficients of u, v, and w in (5.4) are equal to 2 if and only if $a = b = c$; consequently we have equality in the Erdős-Mordell inequality if and only if ABC is equilateral and O is its center.

5.9 The Steiner-Lehmus theorem

It is an easy exercise to prove that, in an isosceles triangle, the angle bisectors of the equal angles are equal in length. The converse is not so easy to prove, and has become known as the *Steiner-Lehmus theorem*. In 1840 C. L. Lehmus asked the Swiss geometer Jakob Steiner (1796–1863) for a purely geometrical proof, as algebraic proofs are relatively simple once one knows the lengths of the angle bisectors in terms of the sides (see [Alsina and Nelsen, 2009]). H. S. M. Coxeter and S. L Greitzer [Coxeter and Greitzer, 1967] write that this result seems "to exert a peculiar fascination on anybody who happens to stumble on" it. Since 1840 hundreds of proofs have been published, and they continue to appear. Our proof is a recent trigonometric proof [Hajja, 2008a] that uses only the law of sines, the double angle sine formula, and the angle bisector theorem (that the bisector of an angle in a triangle divides the side opposite the angle in the same ratio as the sides adjacent to the angle).

The Steiner-Lehmus Theorem 5.12. *If two internal angle bisectors of a triangle are equal, then the triangle is isosceles.*

Proof. Let BB' and CC' be the internal bisectors of angles B and C, and let $a = |BC|$, $b = |AC|$, $c = |AB|$, $B = 2\beta$, $C = 2\gamma$, $u = |AB'|$, $U = |B'C|$, $v = |AC'|$, and $V = |C'B|$, as shown in Figure 5.18.

Assume $|BB'| = |CC'|$ and $C > B$ (and hence $c > b$). We now derive the contradiction that $b/u < c/v$ and $b/u > c/v$.

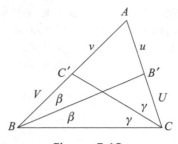

Figure 5.18.

The angle bisector theorem insures that $U/u = a/c$ and $V/v = a/b$, thus

$$\frac{b}{u} - \frac{c}{v} = \frac{u+U}{u} - \frac{v+V}{v} = \frac{U}{u} - \frac{V}{v} = \frac{a}{c} - \frac{a}{b} < 0.$$

The law of sines and the double angle sine formula yield

$$\frac{b}{u} \div \frac{c}{v} = \frac{b}{c}\frac{v}{u} = \frac{\sin B}{\sin C}\frac{v}{u} = \frac{2\cos\beta\sin\beta}{2\cos\gamma\sin\gamma}\frac{v}{u} = \frac{\cos\beta}{\cos\gamma}\frac{\sin\beta}{u}\frac{v}{\sin\gamma}$$

$$= \frac{\cos\beta}{\cos\gamma}\frac{\sin A}{\sin A}\frac{|CC'|}{|BB'|} = \frac{\cos\beta}{\cos\gamma} > 1,$$

yielding the desired conclusion. ■

Given the simplicity of this proof, one might wonder about Coxeter and Greitzer's claim about the "peculiar fascination" with this theorem for some mathematicians. But Lehmus requested a purely geometric proof, which ours clearly is not. In addition, the search has been for direct geometric proofs—nearly all known geometric proofs are, like ours, indirect *reductio ad absurdum* proofs.

5.10 The medians of a triangle

A *median* of a triangle is the line segment joining a vertex to the midpoint of the opposite side. Among some of the remarkable properties of the medians is the following theorem, known to Archimedes. We shall present a modern proof.

Theorem 5.13. *The three medians of a triangle are concurrent, and the point of concurrency (called the* centroid *or center of gravity of the triangle) divides each median into two segments in the ratio* 2:1.

Proof [Rubinstein, 2003]. In triangle ABC, let BF and CD be two medians, intersecting at the point G (see Figure 5.19a). Draw AG, and extend it to meet a line drawn parallel to CD through B at H (see Figure 5.19b). Let E be the intersection of BC and AH, and draw the segment CH.

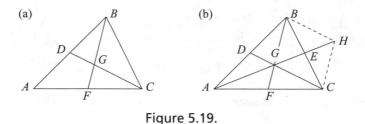

Figure 5.19.

Since D is the midpoint of AB and DG is parallel to BH, G is the midpoint of AH. But F is the midpoint of AC, so that CH is parallel to BF. Hence $BGCH$ is a parallelogram and so its diagonals BC and GH bisect each other. Thus AE is also a median, and the three medians are concurrent at the centroid G. As a bonus, $|AG| = |GH| = 2|GE|$ and similarly for the other two medians, so the centroid divides each median into two segments in the ratio 2:1. ∎

Corollary 5.1. *The three medians of a triangle partition the triangle into six triangles of equal area.*

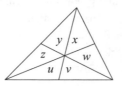

Figure 5.20.

Proof. Let K denote the area of the large triangle, and let u, v, w, x, y, and z denote the areas of the six small triangles, as illustrated in Figure 5.20. Since a median always divides a triangle into two smaller triangles with equal area, we have $u = v$, $w = x$, and $y = z$. Furthermore, $u + v + w = K/2 = v + w + x$, so that $u = x$. Similarly, $v = y$ and $w = z$, and thus $u = v = y = z = w = x$. ∎

Given the lengths of the three sides of a triangle, it is easy to compute the lengths of the three medians (see Challenge 5.7). Furthermore, the three medians can be used to construct a triangle, and such a triangle always has 3/4 the area of the original triangle (see Theorem 10.4).

It is trivial to partition a given triangle into four similar triangles—simply draw lines joining the midpoints of the sides. It is also easy to see that it is impossible to partition a triangle into two, three, or five similar triangles. However, it is amusing to note that all other values are possible, as we show in the next theorem.

Theorem 5.14. *For any $n \geq 6$ a triangle can be partitioned into n similar triangles.*

Proof. Our proof consists of two parts: (a) first we show it true for $n = 6, 7, 8$; and then (b) show that if it is true for $n = k$, it must be true for $n = k + 3$. See Figure 5.21. ∎

$n=6$ $n=7$ $n=8$ $n=k$ $n=k+3$

(a) (b)

Figure 5.21.

The same result is true for squares—see Section 8.4.

5.11 Are most triangles obtuse?

If one chooses three points in the plane "at random," what is the probability that they are the vertices on an acute triangle? a right triangle? an obtuse triangle?

This is an old problem, with multiple solutions. One of the earliest solutions appears to be by Lewis Carroll, best known as the author of *Alice's Adventures in Wonderland* and *Through the Looking-Glass*. Carroll's real name was Charles Lutwidge Dodgson (1832–1898) and he was a lecturer in mathematics at Christ Church, Oxford. In that role he published *Pillow-Problems Thought Out During Wakeful Hours* [Carroll, 1958] in 1893. Problem 58 in that collection reads

> Three Points are taken at random on an infinite Plane. Find the chance of their being the vertices of an obtuse-angled Triangle.

Carroll computes the chance as follows. Let the longest side of the triangle be the segment AB, and without loss of generality let $|AB| = 1$. Then the third vertex C must lie in the intersection of two circles or radius 1 with centers at A and B, i.e., the lens-shaped shaded region in Figure 5.22a. Triangle ABC will be obtuse whenever C lies inside the circle with diameter AB, and hence the probability that ABC is obtuse is the ratio of the darker gray circle to the entire shaded lens-shaped region. This ratio is easily computed, and is

$$\frac{\pi/4}{2\pi/3 - \sqrt{3}/2} \approx 0.639.$$

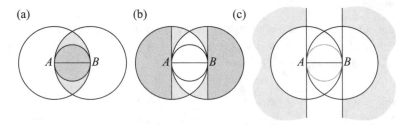

Figure 5.22.

However, if we choose AB to be the second-longest side, then C must lie inside one of the larger circles and outside the other (the shaded region illustrated in Figure 5.22b); and ABC will be obtuse whenever C lies outside the lines perpendicular to AB at A and B. In this case the ratio of areas is

$$\frac{\pi/2}{\pi/3 + \sqrt{3}/2} \approx 0.821.$$

Finally, if we choose AB to be the shortest side, then C must lie outside both large circles and outside of the two lines perpendicular to AB (see Figure 5.22c), and hence the probability that ABC is obtuse is arbitrarily close to 1.

We have obtained three different values for the probability since choosing points "at random in the plane" is not well defined. For discussions and alternate approaches to the problem, see [Guy, 1993; Portnoy, 1994].

5.12   Challenges

5.1 In the classic motion picture *The Wizard of Oz* (1936), the Scarecrow, misstating the Pythagorean theorem, says: "The sum of the square roots of any two sides of an isosceles triangle is equal to the square root of the remaining side." Do any triangles, isosceles or otherwise, satisfy this property?

5.2 Give an algebraic proof of Theorem 5.5.

5.3 Squares have been inscribed in congruent isosceles right triangles in two ways (see Figure 5.23). Which square has the larger area?

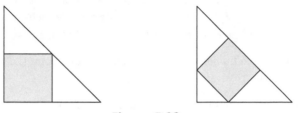

Figure 5.23.

5.4 Prove the angle bisector theorem.

5.5 In triangle ABC, let h_a, h_a, h_a denote the altitudes to sides a, b, c, respectively; and R and r the circumradius and inradius of ABC, respectively. Show that

(a) $h_a + h_b + h_c = \dfrac{ab + bc + ca}{2R}$ and   (b) $\dfrac{1}{h_a} + \dfrac{1}{h_b} + \dfrac{1}{h_c} = \dfrac{1}{r}$.

[Hint: the area K of ABC is given by $K = ah_a/2 = bh_b/2 = ch_c/2$.]

5.6 Inscribing circles, arcs of circles, and drawing lines in a square creates triangles, as seen in Figure 5.24. Prove that the shaded triangle in each part of the figure is similar to the 3:4:5 right triangle [Bankoff and Trigg, 1974].

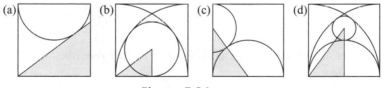

Figure 5.24.

5.7 In triangle ABC, let m_a denote the length of the median from vertex A to the midpoint of the opposite side (of length a); and similarly for m_b and m_c.

(a) Show that $m_a^2 = (2b^2 + 2c^2 - a^2)/4$, similarly for m_b^2 and m_c^2.

(b) Show that $a \leq b \leq c$ implies $m_a \geq m_b \geq m_c$.

5.8 Prove *Carnot's theorem*: In any acute triangle, the sum of the distances from the circumcenter to the sides is equal to the sum of the inradius and the circumradius.

5.9 Prove that the centroid G, the circumcenter O, and the orthocenter H (the intersection of the three altitudes) of a triangle lie on a straight line (known as the *Euler line* of the triangle), and that $|GH| = 2|GO|$.

5.10 Prove that the altitudes of a triangle intersect in a point (i.e., the orthocenter *exists*).

5.11 Given a circle with diameter AB and a point P outside the circle, is it possible to draw a line from P perpendicular to the line determined by AB using only an unmarked straightedge?

5.12 Prove that if the sides of a triangle are in arithmetic progression, then the line through the centroid and the incenter is parallel to one of the sides of the triangle.

5.13 In an isosceles right triangle draw two rays from the right-angled vertex to the hypotenuse, cutting the hypotenuse into three segments as shown in Figure 5.25. Prove that the three segments form a right triangle if and only if the angle between the two rays is 45°.

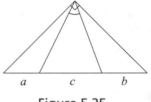

Figure 5.25.

5.14 Prove that Heron's formula and the Pythagorean theorem are equivalent.

CHAPTER 6

The Enchantment of the Equilateral Triangle

> *Of all the figures in plane geometry, the triangle is the most interesting, and the most prolific in terms of producing theorems. Moreover, of all the triangles, the equilateral seems to stand out as perfection personified.*
>
> J. Garfunkel and S. Stahl

Equilateral triangles lie at the heart of plane geometry. In fact Euclid's first proposition—Proposition 1 in Book I of the *Elements*—reads [Joyce, 1996]: *To construct an equilateral triangle on a given finite straight line.* Equilateral triangles continue to fascinate professional and amateur mathematicians. Many of the theorems about equilateral triangles are striking in their beauty and simplicity.

Mathematicians strive to find beautiful proofs for beautiful theorems. In this chapter we present a small selection of theorems about equilateral triangles and their proofs.

6.1 Pythagorean-like theorems

The Pythagorean theorem is usually illustrated with squares on the legs and hypotenuse of the triangle, and many lovely visual proofs employ such illustrations. See Section 5.1 for several examples. However, as a consequence of Proposition 31 in Book VI of the *Elements* of Euclid, any set of three similar figures can be used. See Figure 6.1, where in each case the sum of the areas of the shaded figures on the legs equals the area of the unshaded figure on the hypotenuse.

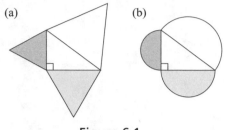

(a)   (b)

Figure 6.1.

We shall use right triangles with semicircles on the sides in Section 9.1 when we explore lunes. In this section we construct equilateral triangles on the three sides of right triangles, as well as on the three sides of acute and obtuse triangles.

Can we prove the Pythagorean theorem with Figure 6.1a? In 1923, the following two problems appeared in the *American Mathematical Monthly*, the first in the July-August issue, the second in the December issue:

3028. *Proposed by Norman Anning, University of Michigan.*

Equilateral triangles are described on the sides of a right triangle. Dissect the triangles on the legs and reassemble the parts to form the triangle on the hypotenuse.

3048. *Proposed by H. C. Bradley, Massachusetts Institute of Technology.*

Cut two equilateral triangles of any relative proportions into not more than five pieces which can be assembled to form a single equilateral triangle.

Solutions by H. C. Bradley to both problems appeared seven years later in 1930 [Bradley, 1930]—an unusually long time to wait for the solution to a *Monthly* problem! In his solutions, Bradley dissects the equilateral triangle on the longer of the two legs and reassembles the pieces, along with the equilateral triangle from the shorter leg, to form the equilateral triangle on the hypotenuse, as illustrated in Figure 6.2.

Other dissections are possible; see [Frederickson, 1997] for details. For a different proof of the Pythagorean theorem using Figure 6.1a, see Section 7.6.

We now extend this idea, constructing equilateral triangles on the sides of a given arbitrary triangle. Let ABC be an arbitrary triangle with sides

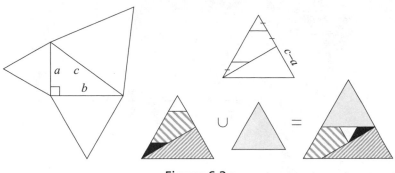

Figure 6.2.

a, b, and c (opposite vertices A, B, and C, respectively) and let T denote its area. Also let T_s denote the area of an equilateral triangle with side s. See Figure 6.3.

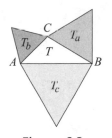

Figure 6.3.

The version of the Pythagorean theorem illustrated in Figure 6.1a can now be stated as: If $C = 90°$, then $T_c = T_a + T_b$. However, we have surprisingly similar results relating T, T_a, T_b, and T_c when C is 60° or 120°.

Theorem 6.1.

(a) *If $C = 60°$, then $T + T_c = T_a + T_b$;*

(b) *if $C = 120°$, then $T_c = T_a + T_b + T$.*

Proof. For (a), we compute the area of an equilateral triangle with sides $a + b$ in two ways to conclude that $3T + T_c = 2T + T_a + T_b$, from which the result follows [Moran Cabre, 2003]. See Figure 6.4.

For (b), we compute in two ways the area of an equiangular hexagon whose sides alternate in length between a and b to conclude that $3T + T_c = 4T + T_a + T_b$, which proves the result [Nelsen, 2004]. See Figure 6.5. ∎

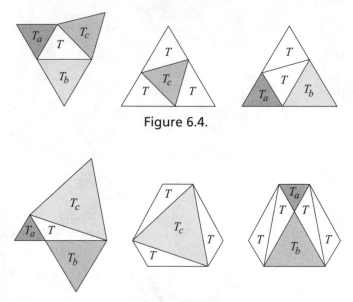

Figure 6.4.

Figure 6.5.

In general, $T_c = T_a + T_b - \sqrt{3}\cot C \cdot T$ (see Challenge 6.1), and as a consequence we have

Corollary 6.1.

(a) If $C = 30°$, then $3T + T_c = T_a + T_b$;

(b) if $C = 150°$, then $T_c = T_a + T_b + 3T$.

6.2   The Fermat point of a triangle

The celebrated mathematician Pierre de Fermat (1601–1665) gave the following problem to Evangelista Torricelli (1608–1647), and Torricelli solved it in several ways:

> Find the point F in (or on) a given triangle ABC such that the sum $|FA| + |FB| + |FC|$ is a minimum. [The point F is now called the *Fermat point* of the triangle.] See Figure 6.6a.

When one of the angles measures 120° or more, the vertex of that obtuse angle is the Fermat point (see Challenge 6.3). So we will consider only triangles in which each angle measures less than 120°. There is a simple way

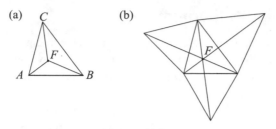

Figure 6.6.

to locate the Fermat point of such a triangle. Construct equilateral triangles on the sides of ABC, and join each vertex of ABC to the exterior vertex of the opposite equilateral triangle. Those three lines intersect at the Fermat point! See Figure 6.6b.

There are many proofs. The one we present was published by J. E. Hofmann in 1929, but it was not new at that time, having been found earlier by Tibor Gallai and others independently [Honsberger, 1973].

Take any point P inside ABC, connect it to the three vertices, and rotate the triangle ABP (shaded in Figure 6.7) 60° counterclockwise as shown to triangle $C'BP'$, and join P' to P.

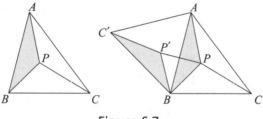

Figure 6.7.

Because triangle $BP'P$ is equilateral, $|AP| = |C'P'|$ and $|BP| = |P'P|$ and so $|AP| + |BP| + |CP| = |C'P'| + |P'P| + |PC|$. Thus the sum $|AP| + |BP| + |CP|$ will be a minimum when P and P' lie on the straight line joining C' to C. There is nothing special about side AB and the new vertex C'—we could equally well have rotated BC or AC counterclockwise (or clockwise) about a vertex. Consequently P must also lie on $B'B$ and $A'A$ (not drawn in the figure), and the Fermat point F is P. In addition, each of the six angles at F measures 60° and the lines joining C to C', B to B', and A to A' (shown in Figure 6.6b, but not in 6.7) all have the same length: $|AP| + |BP| + |CP|$.

6.3   Viviani's theorem

The Fermat point of a triangle is the point in or on the triangle that minimizes the sum of the distances to the vertices. What can we say about a point P in or on the triangle that minimizes the sum of the perpendicular distances to the *sides* of the triangle?

We first consider an equilateral triangle. The surprising result in this case is that P can be any point in or on the triangle, since the sum of the distances from an arbitrary point in an equilateral triangle to its sides is constant. This result is known as *Viviani's theorem*, after Vincenzo Viviani (1622–1703).

Viviani's Theorem 6.2. *The sum of the perpendicular distances from a point in or on an equilateral triangle is equal to the altitude of the triangle.*

Proof. All that is needed is Figure 6.8 [Kawasaki, 2005]. ■

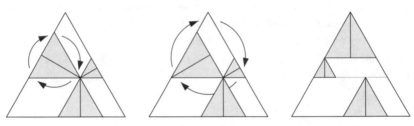

Figure 6.8.

For the case of a non-equilateral triangle, the situation is different. In this case, the point P that minimizes the sum of the distances to the sides is located at the vertex of the largest angle. If there are two largest angles, then there are two locations for P. See Challenge 6.4.

6.4   A triangular tiling of the plane and Weitzenböck's inequality

Multiple copies of the four triangles in Figure 6.3 can be placed as illustrated in Figure 6.9 to *tile the plane*, that is, cover the plane without any overlap. The "tile" is this tiling is the *parahexagon* (a hexagon with opposite sides parallel and equal in length) shown at the right, composed of the three equilateral triangles and three copies of the original (arbitrary) triangle.

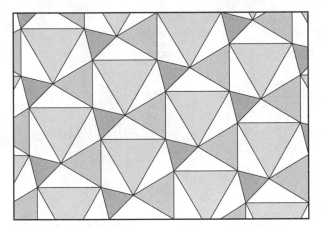

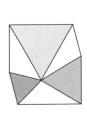

Figure 6.9.

Problem 2 on the Third International Mathematical Olympiad in 1961 read

Let a, b, c be the sides of a triangle, and T its area. Prove

$$a^2 + b^2 + c^2 \geq 4\sqrt{3}T.$$

In what case does equality hold?

This inequality is well known in the problems literature as *Weitzenböck's inequality*, from a paper published in *Mathematische Zeitschrift* in 1919 by R. Weitzenböck. Many analytical proofs are known; in fact there are eleven different proofs in [Engel, 1998]. However, it is not so well known that there is a nice geometric interpretation of the inequality. If we multiply both sides of the inequality by $\sqrt{3}/4$, then it becomes

$$T_a + T_b + T_c \geq 3T, \tag{6.1}$$

that is, at least half the parahexagonal tile in Figure 6.9 is shaded—and thus half the tiling of the plane is shaded—regardless of the shape of the unshaded triangles.

To prove (6.1), we first consider the case where each angle of the given triangle is less than 120°. For such a triangle ABC, let x, y, and z denote the lengths of the line segments joining the Fermat point F to the vertices, as illustrated in Figure 6.10a, and note that the two acute angles in each triangle with a vertex at F sum to 60°.

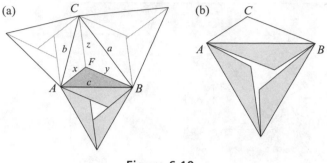

Figure 6.10.

Hence the equilateral triangle with area T_c is the union of three triangles congruent to the dark gray shaded triangle with side lengths x, y, and c, and an equilateral triangle with side length $|x - y|$. Similar results are true for the triangles with areas T_a and T_b, and thus

$$T_a + T_b + T_c = 3T + T_{|x-y|} + T_{|y-z|} + T_{|z-x|}, \qquad (6.2)$$

which establishes (6.1) in this case since $T_{|x-y|}$, $T_{|y-z|}$, and $T_{|z-x|}$ are each nonnegative. We have equality in (6.1) if and only if $x = y = z$, so that the three triangles with a common vertex at F are congruent and hence $a = b = c$, i.e., triangle ABC is equilateral.

When one angle (say C) measures $120°$ or more, then as illustrated in Figure 6.1b we have

$$T_a + T_b + T_c \geq T_c \geq 3T,$$

which completes the proof.

The relationship in (6.2) is actually stronger than the Weitzenböck inequality, and enables us to prove the *Hadwiger-Finsler inequality*: If a, b, and c are the sides of a triangle with area T, then

$$a^2 + b^2 + c^2 \geq 4\sqrt{3}T + (a - b)^2 + (b - c)^2 + (c - a)^2.$$

In terms of areas of triangles, this is equivalent to

$$T_a + T_b + T_c \geq 3T + T_{|a-b|} + T_{|b-c|} + T_{|c-a|}.$$

For a proof, see [Alsina and Nelsen, 2008, 2009].

6.5 Napoleon's theorem

We now return to the tiling of the plane illustrated in Figure 6.9. If the tiling is rotated by 120° clockwise or counterclockwise about the center of any equilateral triangle, it coincides with itself. This rotational symmetry is sufficient to prove another surprising result.

Napoleon's Theorem 6.3. *If equilateral triangles are constructed outwardly on the sides of any triangle, then their centers are the vertices of another equilateral triangle.* See Figure 6.11a.

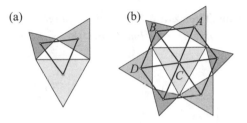

Figure 6.11.

Proof. A portion of the tiling from Figure 6.9 is illustrated in Figure 6.11b. A 120° clockwise rotation of the tiling about B yields $AB = BD$ and an angle of 120° between the segments. Similarly a 120° counterclockwise rotation of the tiling about C yields $AC = CD$ and an angle of 120° between those segments. Thus triangles ABC and BCD are congruent, so BC bisects both $\angle ABD$ and $\angle ACD$, and hence triangle ABC is equilateral. ∎

It is not known whether Napoleon Bonaparte proved this theorem, or if he was even aware of it. However, Napoleon did possess a certain skill in mathematics, and many believe he could have proved it.

There are many other known results about the configuration in Figure 6.11a; we mention only one. Let T_N denote the area of the *Napoleon triangle* whose vertices are the centers of the three original equilateral triangles. Then Figure 6.11b shows that $6T_N = 3T + T_a + T_b + T_c$, or the regular hexagon in Figure 6.11b has the same area as the parahexagonal tile in Figure 6.9. Thus

$$T_N = \frac{1}{2}\left(T + \frac{T_a + T_b + T_c}{3}\right).$$

Since $6(T_N - T) = T_a + T_b + T_c - 3T$, the inequality $T_N \geq T$ is equivalent to Weitzenböck's inequality.

6.6 Morley's miracle

In 1899, Frank Morley (1860–1937) discovered a surprising theorem about triangles that has become known as *Morley's miracle*. The theorem is miraculous not for being deep or powerful, but for never having been discovered before. This may be a result of the fact that trisection of general angles is impossible with straightedge and compass, hence properties of angle trisectors remained unstudied.

Morley's Trisector Theorem 6.4. *The three points of intersection of adjacent angle trisectors in any triangle are the vertices of an equilateral triangle.*

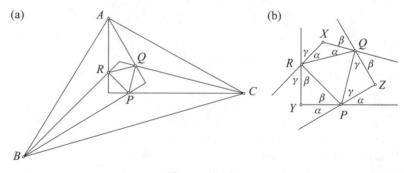

Figure 6.12.

In Figure 6.12a, the trisectors AR, AQ, BP, BR, CP, and CQ intersect to form the equilateral triangle PQR.

Many proofs of Morley's theorem are known, most geometric or trigonometric. However, direct proofs are difficult to construct, so it is easier to work backwards, beginning with an equilateral triangle, and constructing a general triangle similar to the given triangle ABC. Our proof is from [Coxeter, 1961], based on a proof published by Raoul Bricard in 1922.

Proof. See Figure 6.12b. On the sides of an equilateral triangle PQR, construct isosceles triangles RQX, QPZ, and RPY, with base angles α, β, γ that satisfy

$$\alpha + \beta + \gamma = 120°, \quad \alpha < 60°, \quad \beta < 60°, \quad \gamma < 60°.$$

Extend the sides of the three isosceles triangles beyond their bases until they meet in points A, B, and C. Since $\alpha + \beta + \gamma + 60° = 180°$, we can label some of the other angles as indicated in Figure 6.12b (the measures of these angles ensure that the extended sides of the isosceles triangle do indeed meet in points A, B, C as claimed).

Referring to Figure 5.11a, we see that one way to characterize the incenter I of a triangle ABC is to observe that it lies on the bisector of angle A at a distance from A such that

$$\angle BIC = 90° + (1/2)\angle BAC.$$

Using this result for point P in triangle BXC, observe that the line PX (not drawn in the figure, but it is a median of both the isosceles triangle RQX and the equilateral triangle PQR) bisects the angle at X. Since $\angle BXC = \angle RXQ = 180° - 2\alpha$ and $\angle BPC = 180 - \alpha = 90° + (1/2)\angle BXC$, P is the incenter of triangle BXC. Similarly, Q is the incenter of triangle AYC and R is the incenter of triangle AZB. Thus the three small angles at A are equal; similarly at B and C. Thus the angles of ABC are trisected.

The three small angles at A are each $(1/3)\angle A = 60° - \alpha$ (likewise at B and C), and thus

$$\alpha = 60° - (1/3)\angle A, \quad \beta = 60° - (1/3)\angle B, \quad \gamma = 60° - (1/3)\angle C;$$

so that by choosing these values for the base angles α, β, γ of the isosceles triangles RQX, QPZ, and RPY, the procedure yields a triangle ABC similar to any given triangle. ∎

The Sierpiński triangle

One of the prettiest fractals is *Sierpiński's triangle*, also known as the Sierpiński sieve or gasket. It was first described by the Polish mathematician Waclaw Sierpiński in 1915. To construct it, begin with an equilateral triangle and delete the central one-fourth, then delete the central one-fourth of each of the remaining smaller equilateral triangles, and continue the process indefinitely. The first few steps in this process are shown in Figure 6.13.

Figure 6.13.

Fractals are self-similar sets. The Sierpiński triangle consists of three copies of itself, each scaled by a factor of $1/2$. As a consequence, its Hausdorff dimension is $\ln 3 / \ln 2 \approx 1.585$.

6.7 Van Schooten's theorem

The following simple but surprising theorem is attributed to the Dutch mathematician Franciscus Van Schooten (1615–1660). Many proofs of Van Schooten's theorem are possible; our proof parallels the one we used to locate the Fermat point of a triangle.

Van Schooten's Theorem 6.5. *An equilateral triangle ABC is inscribed in a circle. Let P be any point on the circle, and draw the segments AP, BP, and CP. The length of the longest of the three segments equals the sum of the lengths of the shorter two.*

Proof. See Figure 6.14a. We claim that $|AP| = |BP| + |CP|$. First rotate triangle ABC and the three segments from P to the vertices $60°$ clockwise about vertex B to obtain the configuration in Figure 6.14b. Let C' and

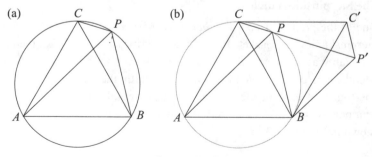

Figure 6.14.

P' denote the images of C and P in the rotation. Now $BP = BP'$ and $\angle PBP' = 60°$, so triangle BPP' is equilateral. But $\angle CPB = 120°$, so the points C, P, and P' lie on a line. Hence $|AP| = |CP'| = |CP| + |PP'| = |CP| + |BP|$. ■

Curves of constant width and the Reuleaux triangle

The *width* of a convex closed curve is the distance between two parallel lines lying outside the curve but touching its boundary. If the width is the same for every such pair of parallel lines, then it is a *curve of constant width*. The circle is a curve of constant width, but so is the *Reuleaux triangle* (Franz Reuleaux, 1829–1905). To construct it, draw circular arcs using each vertex of an equilateral triangle as the center and the side of the triangle as the radius. See Figure 6.15.

Figure 6.15.

Of all curves of constant width w, the one with the largest area is the circle; the one with the smallest area is the Reuleaux triangle. Somewhat surprising is the fact that both have the same perimeter, πw. Even more surprising is *Barbier's theorem* (Joseph Emile Barbier, 1839–1889): *Every* convex curve with constant width w has the same perimeter πw! For a proof, see [Honsberger, 1970].

6.8 The equilateral triangle and the golden ratio

As we noted in Section 2.3, there is a natural relationship between the golden ratio and the regular pentagon. It is not so well known that the golden ratio also makes an appearance in the equilateral triangle and its circumcircle. The following problem appeared in the August-September 1983 issue of the *American Mathematical Monthly*.

E 3007. *Proposed by George Odom, Poughkeepsie, NY.*

Let A and B be the midpoints of the sides EF and ED of an equilateral triangle DEF. Extend AB to meet the circumcircle (of DEF) at C. Show that B divides AC according to the golden section.

A solution [van de Craats, 1986] appeared three years later, as Figure 6.16a. Since the shaded triangles are similar, $(1 + x)/x = x/1$, so $x^2 = x + 1$. Since x is positive, it is the golden ratio φ.

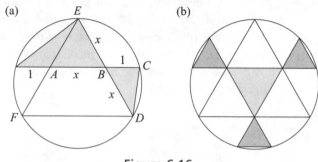

Figure 6.16.

This result leads to the attractive design in Figure 6.16b, where the ratio of the side of a large shaded triangle to that of the small shaded triangle is φ, and the corresponding ratio of areas is $\varphi + 1$ [Rigby, 1988].

6.9   Challenges

6.1 Using the notation in Section 6.1, prove that $T_c = T_a + T_b - \sqrt{3} \cot C \cdot T$.

6.2 Create a visual proof of Corollary 6.1 analogous to those in Figures 6.4 and 6.5.

6.3 Prove that if one angle of a triangle measures $120°$ or more, then it is the Fermat point of the triangle. [Hint: See Figure 6.17.]

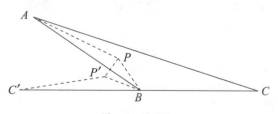

Figure 6.17.

6.4 Prove that for a non-equilateral triangle, the point P in or on the triangle that minimizes the sum of the perpendicular distances from P to the sides is the vertex of the largest angle.

6.5 Show that every triangle has infinitely many inscribed equilateral triangles.

6.6 Use the equilateral triangles in Figure 6.18 to show that $\sqrt{3}$ is irrational.

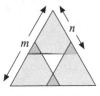

Figure 6.18.

6.7 Prove *Pompeiu's theorem* (Dimitrie Pompeiu, 1873–1954): Let ABC be an equilateral triangle, and P any point in the plane of ABC. Then the lengths $|AP|, |BP|, |CP|$ always form a (possibly degenerate) triangle.

6.8 In square $ABCD$ locate point E as shown in Figure 6.19. Prove that triangle ABE is equilateral.

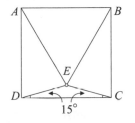

Figure 6.19.

6.9 It is easy to show that the area of an isosceles right triangle is one-fourth of the square of the hypotenuse (consider a square and its two diagonals). Prove that the area of a right triangle with a 15° angle is one-eighth of the square of the hypotenuse. [Hint: use the preceding Challenge.]

6.10 Inscribe an equilateral triangle in a square as shown in Figure 6.20a, with one side making 45° angles with sides of the square. Prove that (i) the sum of the areas of the two dark gray triangles in Figure 6.20b equals the area of the light gray triangle; and (ii) if small isosceles right triangles are removed from the dark gray triangles as shown in Figure 6.20c, then the sum of the areas of the three shaded triangles equals the area of the white equilateral triangle.

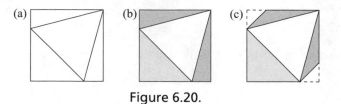

Figure 6.20.

6.11 Draw equilateral triangles ABE and ADF inside and outside, respectively, of square $ABCD$ as shown in Figure 6.21. Prove that points C, E, and F are collinear.

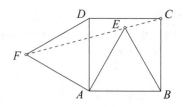

Figure 6.21.

CHAPTER 7

The Quadrilaterals' Corner

> *The landlady of a boarding-house is a parallelogram—that is, an oblong angular figure, which cannot be described, but which is equal to anything.*
>
> Stephen Leacock
> *Boarding-House Geometry* (1910)

Euclid's *Elements* contains approximately three dozen propositions concerning properties of triangles, but only about a dozen concerning properties of quadrilaterals, and most of those deal with parallelograms. These statistics belie the richness found in the set of quadrilaterals and its various subsets: cyclic, bicentric, parallelograms, trapezoids, squares, and so on. In this chapter we discuss some intriguing results and their proofs concerning general quadrilaterals as well as some of the special cases just mentioned.

Triangles may be acute, right, or obtuse and equilateral, isosceles, or scalene. Similarly, quadrilaterals may be *planar* or *skew* (non-planar); planar quadrilaterals may be *complex* (self-intersecting) or *simple* (non-self-intersecting); and simple quadrilaterals may be *convex* (each interior angle less than 180°) or *concave* (one interior angle greater than 180°).

7.1   Midpoints in quadrilaterals

Our first theorem is a remarkable result that applies to any quadrilateral—convex, concave, complex, or skew. It was first published by Pierre Varignon (1654–1722) and is known as *Varignon's theorem*.

Varignon's Theorem 7.1. *The midpoints of the sides of an arbitrary quadrilateral form a parallelogram.*

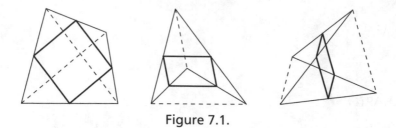

Figure 7.1.

Proof. See Figure 7.1. In each case, the line segment joining the midpoints of adjacent sides is parallel to and half the length of one of the diagonals as a consequence of Proposition VI.2 in Euclid's *Elements*. Hence the midpoint quadrilateral has opposite sides parallel and is thus a parallelogram. ∎

The parallelogram in Theorem 7.1 in known as the *Varignon parallelogram* associated with the given quadrilateral. The two *bimedians* (the lines joining the midpoints of opposite sides) of the quadrilateral are the diagonals of the Varignon parallelogram, and hence bisect each other.

It follows immediately from the proof of Theorem 7.1 that the perimeter of the Varignon parallelogram is equal to one-half the sum of the lengths of the diagonals of the quadrilateral. It is also easy to show (in the convex case) that the area of the Varignon parallelogram is equal to one-half the area of the quadrilateral (see Challenge 7.1).

Perhaps more surprising than Varignon's theorem is the following result.

Theorem 7.2. *In any convex quadrilateral, the point of intersection of the bimedians bisects the line segment joining the midpoints of the diagonals. See Figure 7.2a.*

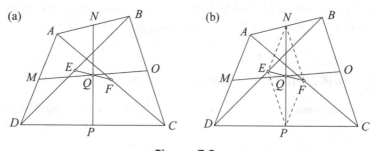

Figure 7.2.

Proof. Since the bimedians of the quadrilateral $ABCD$ bisect each other, Q is the midpoint of both MO and NP (see Figure 7.2b). Since EN and FP

are each parallel to AD, and EP and FN are each parallel to BC, $ENFP$ is a parallelogram. Hence its diagonals EF and NP bisect each other, and thus Q is the midpoint of EF, as claimed. ∎

The line joining the midpoints of the diagonals of a quadrilateral that is not a parallelogram (EF in Figure 7.2a) is called the *Newton line* of the quadrilateral, and plays an important role in several of the results to come in this chapter.

We conclude this section with two more nice area properties of convex quadrilaterals. For the first, draw lines from opposite vertices to midpoints of opposite sides, as illustrated in Figure 7.3a. This creates a smaller shaded quadrilateral within the original one, and its area is one-half the area of the original, as illustrated in Figure 7.3b.

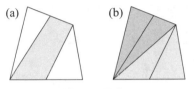

Figure 7.3.

For the second, let RS be any segment on the Newton line EF of the quadrilateral $ABCD$ (see Figure 7.4), and form two triangles using RS as the base and opposite vertices as the third vertex (e.g., A and C, or B and D). Then the two triangles have the same area, since they have the same base and altitude.

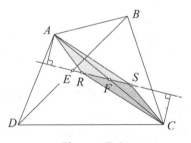

Figure 7.4.

7.2 Cyclic quadrilaterals

A *cyclic* quadrilateral is one whose four vertices lie on a circle. As a consequence of Proposition 22 in Book III of Euclid's *Elements*, the angles at

opposite vertices of a cyclic quadrilateral are supplementary, i.e., they sum
to 180°.

In this section we show that among all simple quadrilaterals with given
side lengths, the one with the largest area is a cyclic quadrilateral. Our proof
uses nothing more than trigonometric identities and algebra. First we present
a lemma that answers the following question: given any simple quadrilateral,
does there exist a cyclic quadrilateral with the same side lengths?

Lemma 7.1. *For any simple quadrilateral with given side lengths, there is
a cyclic quadrilateral with the same side lengths.*

Proof. [Peter, 2003]. Let Q be a simple quadrilateral with sides a, b, c, and
d as illustrated in Figure 7.5. We can assume that Q is convex since every
concave quadrilateral is contained in a convex quadrilateral with the same
side lengths. Assume that the sides are labeled so that $a + b \leq c + d$.

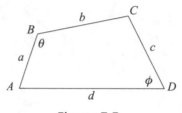

Figure 7.5.

Now think of Q as having hinges at the four vertices. By pushing on the
hinge at B, we can deform the quadrilateral into a triangle with A, B, and
C collinear. In this triangle $\theta = \pi$ and $\phi > 0$ so that we have $\theta + \phi > \pi$.
Without loss of generality we can also assume that $b + c \leq a + d$. Pulling
on the hinge at B deforms Q into a triangle with B, C, and D collinear, and
in this triangle we have $\theta + \phi < \pi$. Hence we can find a position for B
where $\theta + \phi = \pi$, and this quadrilateral is cyclic. ■

Theorem 7.3. *The cyclic quadrilateral Q has the largest area of all quadrilaterals with sides the same length as those of Q.*

Proof. See Figure 7.5. Applying the law of cosines in triangles ABC and
ACD yields

$$a^2 + b^2 - 2ab \cos \theta = c^2 + d^2 - 2cd \cos \phi$$

with θ, ϕ in $(0,\pi)$, so that

$$0 = 4(a^2 b^2 \cos^2 \theta - 2abcd \cos \theta \cos \phi + c^2 d^2 \cos^2 \phi) - (a^2 + b^2 - c^2 - d^2)^2.$$
$$(7.1)$$

The area K of Q is the sum of the areas of triangles ABC and ACD, so that $K = (1/2)(ab \sin \theta + cd \sin \phi)$ and thus

$$16K^2 = 4a^2b^2 \sin^2 \theta + 8abcd \sin \theta \sin \phi + 4c^2d^2 \sin^2 \phi. \qquad (7.2)$$

Adding (7.1) to (7.2) and simplifying yields

$$16K^2 = 4(a^2b^2 + c^2d^2) - (a^2 + b^2 - c^2 - d^2)^2 - 8abcd \cos(\theta + \phi).$$
$$(7.3)$$

Hence $16K^2$ (and consequently K) is maximized when $\cos(\theta + \phi)$ is a minimum, which occurs when $\theta + \phi = \pi$, i.e., when Q is cyclic. ∎

The above proof also yields a nice expression for the area of a cyclic quadrilateral, known as *Brahmagupta's formula*. When Q is cyclic, we have

$$\begin{aligned}
16K^2 &= 4(a^2b^2 + c^2d^2) - (a^2 + b^2 - c^2 - d^2)^2 + 8abcd \\
&= 4(ab + cd)^2 - (a^2 + b^2 - c^2 - d^2)^2 \\
&= [(a + b)^2 - (c - d)^2][(c + d)^2 - (a - b)^2] \\
&= (a + b + c - d)(a + b - c + d)(a - b + c + d)(-a + b + c + d),
\end{aligned}$$

which proves

Corollary 7.1 (Brahmagupta's formula). *The area K of the cyclic quadrilateral Q with side lengths a, b, c, and d is*

$$K = \sqrt{(s - a)(s - b)(s - c)(s - d)},$$

where $s = (a + b + c + d)/2$ is the semi-perimeter of Q.

We can also prove Theorem 7.3 using calculus [Peter, 2003] or the isoperimetric theorem [Alsina and Nelsen, 2009].

7.3 Quadrilateral equalities and inequalities

There is a wealth of lovely equalities and inequalities for a quadrilateral Q with area K, sides a, b, c, d, diagonals p and q, and if Q is cyclic, circumradius R. We begin with *Ptolemy's theorem* for cyclic quadrilaterals. It is generally credited to Claudius Ptolemy of Alexandria (circa 85–165). There are many proofs of Ptolemy's theorem, but perhaps the nicest is the one presented by Ptolemy himself in the *Almagest*, which we now present.

Ptolemy's Theorem 7.4. *In a cyclic quadrilateral Q, the product of the diagonals equals the sum of the products of the two pairs of opposite sides, i.e., if Q has sides a, b, c, d (in that order) and diagonals p and q, then* $pq = ac + bd$.

Proof. Figure 7.6a illustrates a cyclic quadrilateral $ABCD$, and in Figure 7.6b we choose the point E on the diagonal BD and draw the segment CE so that $\angle BCA = \angle DCE$. Let $|BE| = x$ and $|ED| = y$, with $x + y = q$.

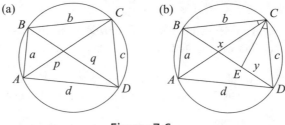

Figure 7.6.

The angles marked ∡ in Figure 7.7a subtend the same arc, hence they are equal, and the shaded triangles are similar. Thus $a/p = y/c$, or $ac = py$. Similarly in Figure 7.7b the shaded triangles are similar, and $d/p = x/b$, or $bd = px$.

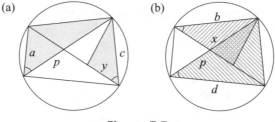

Figure 7.7.

Thus $ac + bd = p(x + y) = pq$. ■

Ptolemy's theorem can be strengthened to *Ptolemy's inequality* for convex quadrilaterals: *If Q is a convex quadrilateral with sides a, b, c, d (in that order) and diagonals p and q, then* $pq \leq ac + bd$, *with equality if and only if Q is cyclic.* See [Alsina and Nelsen, 2009] for a proof.

The following theorem, which complements Ptolemy's theorem, yields an expression for the ratio of the diagonals in a cyclic quadrilateral, and consequently expressions for the lengths of the diagonals in terms of the sides.

As a bonus we obtain an expression for the area of a cyclic quadrilateral in terms of the sides and circumradius.

Theorem 7.5. *Let Q be a cyclic quadrilateral with sides and diagonals as indicated in Figure 7.6a, area K and circumradius R. Then*

(a) $\dfrac{p}{q} = \dfrac{ad + bc}{ab + cd}$,

(b) $p = \sqrt{\dfrac{(ac + bd)(ad + bc)}{ab + cd}}$ and $q = \sqrt{\dfrac{(ac + bd)(ab + cd)}{ad + bc}}$,

(c) $4KR = \sqrt{(ab + cd)(ac + bd)(ad + bc)}$,

Proof. Lemma 5.3 yields an expression for the area of a triangle in terms of its three sides and the circumradius, which we now apply to the triangles with areas K_1, K_2, K_3, and K_4 formed by two sides and a diagonal of Q, as illustrated in Figure 7.8.

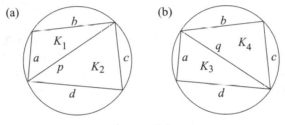

Figure 7.8.

Thus

$$K = K_1 + K_2 = \frac{pab}{4R} + \frac{pcd}{4R} = \frac{p(ab + cd)}{4R},$$

$$K = K_3 + K_4 = \frac{qad}{4R} + \frac{qbc}{4R} = \frac{q(ad + bc)}{4R},$$

hence $p(ab + cd) = q(ad + bc)$, and part (a) follows. For (b) and (c) we have

$$p^2 = pq \cdot \frac{p}{q} = \frac{(ac + bd)(ad + bc)}{ab + cd},$$

$$q^2 = pq \cdot \frac{q}{p} = \frac{(ac + bd)(ab + cd)}{ad + bc},$$

$$K^2 = \frac{pq(ab + cd)(ad + bc)}{(4R)^2} = \frac{(ac + bd)(ab + cd)(ad + bc)}{(4R)^2}.$$

∎

To relate the area K to the diagonals p and q of a convex quadrilateral Q, we have $K \leq pq/2$, with equality if and only if the diagonals of Q are perpendicular (see Challenge 7.5). The following theorem gives similar inequalities for K and the sides a, b, c, and d, whose proof we leave to Challenge 7.7.

Theorem 7.6. *Let Q be a convex quadrilateral with sides as indicated in Figure 7.6a, area K, and perimeter L. Then*

(a) *$K \leq (ab + cd)/2$ and $K \leq (ad + bc)/2$, with equality in one part if two opposite angles of Q are right angles,*

(b) *$K \leq (a + c)(b + d)/4$, with equality if and only if Q is a rectangle,*

(c) *$K \leq L^2/16$, with equality if and only if Q is a square.*

Part (c) of the preceding theorem yields the *isoperimetric inequality for quadrilaterals*: among all quadrilaterals with a fixed perimeter, the one with the largest area is the square.

The presence of the term $ac + bd$ in Theorems 7.4 and 7.5 (as well as $ab + cd$ and $ad + bc$ in Theorems 7.5 and 7.6) suggests another well-known result that applies to quadrilaterals—a special case of the two-dimensional version of the *Cauchy-Schwarz inequality*: for positive numbers a, b, c, and d, we have

$$ac + bd \leq \sqrt{a^2 + b^2}\sqrt{c^2 + d^2}. \tag{7.4}$$

While we shall wait until Section 12.3 to present a one-line proof of the n-dimensional version of the Cauchy-Schwarz inequality, we give a geometric proof using a rectangle partitioned in two different ways [Kung, 2008]:

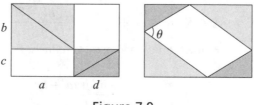

Figure 7.9.

The area of the unshaded region of the left hand rectangle is $ac + bd$, equal to the area $\sqrt{a^2 + b^2}\sqrt{c^2 + d^2}\sin\theta$ of the unshaded parallelogram on the right, which in turn is less than or equal to $\sqrt{a^2 + b^2}\sqrt{c^2 + d^2}$. Figure 7.9 is similar to the *Zhou bi suan jing* proof of the Pythagorean theorem in Figure 5.1.

7.4 Tangential and bicentric quadrilaterals

A quadrilateral is *tangential* if it admits an incircle, and *bicentric* if it is both cyclic and tangential. The following theorems present some properties of tangential and bicentric quadrilaterals.

Theorem 7.7. *Let Q be a tangential quadrilateral with sides a, b, c, d (in that order), where $a = x + y$, $b = y + z$, $c = z + t$, and $d = t + x$, as illustrated in Figure 7.10. Then*

(a) $a + c = b + d$, *and*

(b) Q *is cyclic if and only if $xz = ty$.*

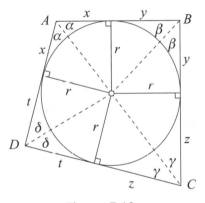

Figure 7.10.

Proof. We can label the sides of the quadrilateral as shown in Figure 7.10 since two line segments tangent to a circle from a point outside the circle have the same length. For (a), note that the sum of either pair of opposite sides is $x + y + z + t$. To prove (b) [Hajja, 2008b], we let 2α, 2β, 2γ, and 2δ denote the angles at vertices A, B, C, and D, respectively, and first show that Q is cyclic if and only if $\tan\alpha \tan\gamma = \tan\beta \tan\delta$. If Q is cyclic, then $\alpha + \gamma = \beta + \delta = \pi/2$, and thus $\tan\alpha \tan\gamma = \tan\beta \tan\delta$ since each side is equal to 1. If Q is not cyclic, then we may assume that $\alpha + \gamma > \pi/2$ and $\beta + \delta < \pi/2$. Since $0 > \tan(\alpha + \gamma) = (\tan\alpha + \tan\delta)/(1 - \tan\alpha \tan\delta)$ and both α and γ are acute, $\tan\alpha \tan\gamma > 1$. Similarly $\tan\beta \tan\delta < 1$, and thus $\tan\alpha \tan\gamma \neq \tan\beta \tan\delta$. The desired result now follows from $\tan\alpha = r/x$, $\tan\beta = r/y$, etc. ∎

Theorem 7.8. *Let Q be a bicentric quadrilateral with sides a, b, c, d (in that order), diagonals p and q, area K, inradius r, and circumradius R. Then*

(a) $K = \sqrt{abcd}$,

(b) $2R^2 \geq K \geq 4r^2$,

(c) $8pq \leq (a+b+c+d)^2$.

Proof. For (a), the semiperimeter s is given by $s = a + c = b + d$, so Brahmagupta's formula (Corollary 4.1) yields $K = \sqrt{abcd}$. The left hand inequality in (b) follows from $K \leq pq/2$ and the fact that each diagonal is less than or equal to $2R$ when Q is cyclic. To see the right hand inequality in (b), observe that when Q is tangential and cyclic, $\tan \alpha \tan \gamma = 1$ and thus $r^2 = xz$. Analogously $r^2 = ty$ and employing the arithmetic mean-geometric mean inequality yields

$$K = r(x+y+z+t) = 2r\left(\frac{x+z}{2} + \frac{t+y}{2}\right) \geq 2r(\sqrt{xz} + \sqrt{ty}) = 4r^2,$$

with equality if and only if $x = y = z = t$, i.e., Q is square. In (c) we use Ptolemy's theorem and the arithmetic mean-geometric mean inequality to obtain

$$8pq = 2(4ac + 4bd) \leq 2[(a+c)^2 + (b+d)^2] = 4s^2 = (a+b+c+d)^2.$$

∎

7.5　Anne's and Newton's theorems

A classical result of Isaac Newton (1642–1727) on tangential quadrilaterals can be easily proved as an immediate consequence of the following result of Pierre-Léon Anne (1806–1850), which is of interest in its own right.

Anne's Theorem 7.9. *Let P be a point inside any convex quadrilateral that is not a parallelogram, and connect P to each of the four vertices. Then the locus of points P such that the sums of the areas of opposite triangles are equal is the Newton line of the quadrilateral, i.e., the line through the midpoints E and F of the diagonals. See Figure 7.11.*

Proof. We need to show that if area(APB) + area(CPD) = area(APD) + area(BPC), then P lies on the line determined by E and F. Introduce a co-ordinate system (with arbitrary origin but axes not parallel to any of the sides

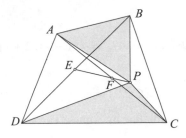

Figure 7.11.

of the quadrilateral), and let $P = (p, q)$. Then the signed perpendicular distance d from P to a fixed line $y = mx + b$ is $d = (q - mp - b)/\sqrt{m^2 + 1}$, a linear function of p and q. Thus the area of a triangle with a fixed base (one of the sides of the quadrilateral) and third vertex P is a linear function of p and q, and the same is true of the sum of the areas of two such triangles. So the locus of points P satisfying the area constraint is a straight line. But area(AEB) = area(AED) and area(CED) = area(BEC) since E is the midpoint of BD, thus E lies on the locus for P. A similar result holds for F, and hence the locus of points P is the line through E and F. ∎

As a consequence we have

Newton's Theorem 7.10. *The center of the circle inscribed in a tangential quadrilateral lies on the Newton line of the quadrilateral.*

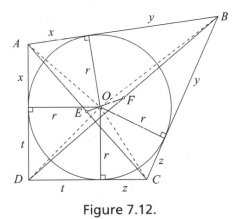

Figure 7.12.

Proof. Label the angles and sides of the quadrilateral as we did in Figure 7.10. Letting r denote the radius of the inscribed circle and O its center, we have

$$r(x + y)/2 + r(t + z)/2 = r(x + t)/2 + r(y + z)/2$$

or

$$\text{area}(AOB) + \text{area}(COD) = \text{area}(AOD) + \text{area}(BOC).$$

Hence O lies on the Newton line by Anne's Theorem. ∎

7.6 Pythagoras with a parallelogram and equilateral triangles

Given a right triangle with legs a and b and area T, it is easy to construct a parallelogram with area P equal to T and angles of 30° and 150°, as shown in Figure 7.13.

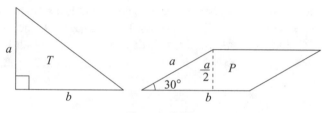

Figure 7.13.

With this construction, we can give a nice proof of the equilateral triangle version of the Pythagorean theorem in Proposition 31 in Book VI of the *Elements* of Euclid. This proof is in a sense a triangular analog of the *Zhou bi suan jing* proof in Section 5.1. We begin by constructing equilateral triangles on the sides of the right triangle as illustrated in Figure 6.1a and below in Figure 7.14a. Let T denote the area of the triangle, P the area of

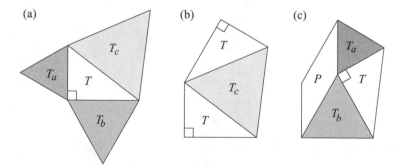

Figure 7.14.

the parallelogram constructed in Figure 7.12 (so that $P = T$), and T_a, T_b, and T_c the areas of the equilateral triangles constructed on the sides a, b, and c of a right triangle. In Figures 7.14b and 7.14c we partition an irregular pentagon in two different ways to conclude that $T_c + 2T = T_a + T_b + P + T$, or equivalently, $T_c = T_a + T_b$.

7.7 Challenges

7.1 Prove that the area of the Varignon parallelogram associated with a convex quadrilateral is one-half the area of the quadrilateral.

7.2 Prove *Bretschneider's formula*: If Q is a simple quadrilateral with sides a, b, c, and d and if θ and ϕ are a pair of opposite angles, then the area K of Q is given by

$$K = \sqrt{(s-a)(s-b)(s-c)(s-d) - abcd\,\cos^2\left((\theta+\phi)/2\right)}.$$

7.3 Use Corollary 7.1 to give another proof of Heron's formula (Theorem 5.7) for the area of a triangle.

7.4 Show that Ptolemy's theorem implies (a) the Pythagorean theorem, (b) van Schooten's theorem 6.5, (c) the length of the diagonal of a regular pentagon with side length 1 is φ, and (d) the addition law for sines: $\sin(\alpha + \beta) = \sin\alpha\cos\beta + \cos\alpha\sin\beta$.

7.5 Prove that the area K of a convex quadrilateral Q with diagonals p and q satisfies $K \le pq/2$, with equality if and only if the diagonals of Q are perpendicular.

7.6 Prove that the points of intersection of the angle bisectors in a convex quadrilateral are the vertices of a cyclic quadrilateral. See Figure 7.15.

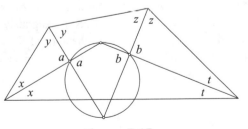

Figure 7.15.

7.7 Prove Theorem 7.6.

7.8 Four circles of the same radius r intersect in a common point, and four
tangents are drawn to form a circumscribing quadrilateral $ABCD$, as
shown in Figure 7.16. Prove that $ABCD$ is cyclic [Honsberger, 1991].

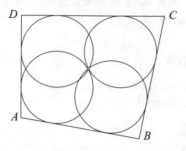

Figure 7.16.

7.9 If we omit the hypothesis that the quadrilateral is not a parallelogram
from Anne's theorem 7.9, how does the locus of the points P change?

7.10 Prove *Miquel's Theorem*: Let Q, R, and S be points chosen arbitrarily
on the sides of triangle ABC (see Figure 7.17). Then the circles ARS,
BSQ, and CQR are concurrent.

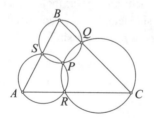

Figure 7.17.

7.11 In Section 6.2 we discussed the Fermat point of a triangle. The *Fermat
point of a convex quadrilateral* is defined analogously as the point in
or on the quadrilateral that minimizes the sum of the distances to the
four vertices. Prove that the Fermat point of a convex quadrilateral is
the intersection of the diagonals.

Squares Everywhere

> *We must say that there are as many squares as there are*
> *numbers.*
>
> Galileo Galilei

Squares have a special place in the world of quadrilaterals, just as equilateral triangles have a special place among all the triangles. We devote this chapter to theorems about squares, both in the geometric and number-theoretic sense. The two are closely related, as you read in Section 3.2 concerning the representation of an integer as the sum of two squares and will see again in Sections 8.2 and 8.3.

We present our theorems about squares according to the number of squares in the theorem. For example, the Pythagorean theorem can be thought of as a three-square theorem.

8.1   One-square theorems

The golden ratio φ appears in many constructions with regular polygons. In Section 2.3 we saw the close relationship between the golden ratio and the regular pentagon, and in Section 6.8 we discovered a relationship between the golden ratio and the equilateral triangle. The following theorem presents a similar result relating the golden ratio and the square.

Theorem 8.1. *Inscribe a square in a semicircle as illustrated in Figure* 8.1a. *Then* $AB/BC = \varphi$.

Proof. See Figure 8.1b. Choose the scale so that $BC = 1$ and let $AB = x$. The shaded triangles are similar, so $x/1 = (x + 1)/x$, and hence $x^2 = x + 1$. Since x is positive, it is the golden ratio φ, and $AB/BC = x = \varphi$. Note that the union of the original square and the dashed rectangle in Figure 8.1 forms a golden rectangle.   ■

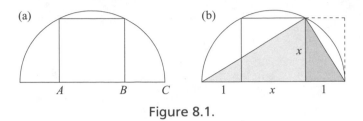

Figure 8.1.

Perhaps the best-known right triangle with integer sides is the 3:4:5 triangle. It can be constructed from a square in the following manner.

Theorem 8.2. *In a square draw line segments from the midpoints of a pair of adjacent sides to opposite vertices as illustrated in Figure 8.2a. The resulting shaded triangle is a triangle similar to the 3:4:5 right triangle.*

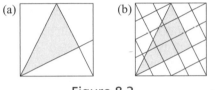

Figure 8.2.

Proof. See Figure 8.2b [DeTemple and Harold, 1996]. ∎

We will have more to say about integer sided right triangles in Section 8.3.

A *root-rectangle* [Hambidge, 1967] is a rectangle in which the ratio of the longer side to the shorter is $\sqrt{n}$ for an integer n. In Figure 8.3 we illustrate an iterative procedure for constructing $\sqrt{n}$-by-1 root-rectangles beginning with a unit (i.e., a 1-by-1) square.

In Figure 8.3, a circular arc is drawn clockwise with center $(0,0)$ and radius $\sqrt{n+1}$ from $(\sqrt{n}, 1)$. Such an arc then intersects the x-axis at $(\sqrt{n+1}, 0)$.

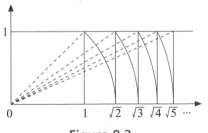

Figure 8.3.

It is also easy to construct a 1-by-$1/\sqrt{n}$ root-rectangles within a unit square with only one circular arc, as shown in Figure 8.4. We generate the figure by finding the intersection of a dashed line (beginning with $y = x$) with the quarter-circle, and drawing a horizontal line through that point. See Challenge 8.1 for the verification that the rectangles do indeed have heights $1/\sqrt{n}$.

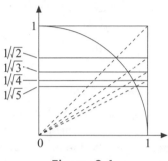

Figure 8.4.

8.2   Two-square theorems

If we inscribe a square in a semicircle (as in Theorem 8.1) and another square in a circle of the same radius, how do the areas of the two squares compare? See Figure 8.5.

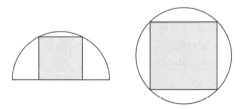

Figure 8.5.

Theorem 8.3. *A square inscribed in a semicircle has* $2/5$ *the area of a square inscribed in a circle of the same radius.*

Proof. Draw a circle with radius $\sqrt{5}$ centered at a lattice point and inscribe a square in the circle and one in the upper semicircle, as shown in Figure 8.6. The area of the smaller square is 4 while the area of the larger is 10 (because its side has length $\sqrt{10}$), hence the ratio of the areas is $2/5$.   ∎

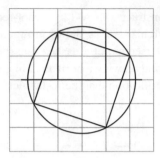

Figure 8.6.

Squares and the measurement of area

Beyond being the second regular polygon, the square plays a key role in mathematics as a tool for the measurement of area. One can speculate that this use of squares underlies the notion of *quadrature*—constructing (with compass and straightedge) a square equal in area to a given figure, also called *squaring* the figure. For example, Proposition 14 in Book II of the *Elements* of Euclid is "To construct a square equal to a given rectilinear figure." One of the three great problems of classical geometry is the problem of squaring a circle.

In the next theorem we use a procedure similar to the one in Theorem 8.2 to construct a square within a square, and compare areas.

Theorem 8.4. *Create a square within a square by drawing a line from the midpoint of each side to one of the opposite vertices, as illustrated in Figure 8.7a. Then the area of the smaller square is 1/5 the area of the larger.*

Proof. See Figure 8.7b. ∎

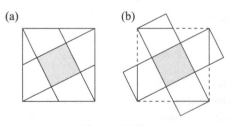

Figure 8.7.

In the next theorem we present a surprising result about two squares sharing a common vertex. It is sometimes called the *Finsler-Hadwiger* theorem. Our proof is from [DeTemple and Harold, 1996].

The Finsler-Hadwiger Theorem 8.5. *Let the squares $ABCD$ and $AB'C'D'$ share a common vertex A, as shown in Figure 8.8. Then the midpoints Q and S of the segments BD' and $B'D$ together with the centers R and T of the original squares form another square $QRST$.*

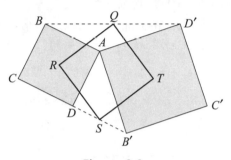

Figure 8.8.

Proof. Construct congruent parallelograms $ABED'$ and $ADFB'$, as shown in Figure 8.9 and observe that Q and S are the centers of $ABED'$ and $ADFB'$. A 90° clockwise rotation about R takes $ABED'$ to $ADFB'$ and the segment RQ to RS. Hence QR and RS are equal in length and perpendicular. A similar 90° counterclockwise rotation about T shows that QT and

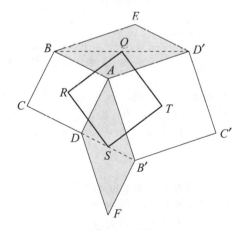

Figure 8.9.

TS are equal in length and perpendicular, from which it follows that $QRST$ is a square. ∎

In Theorem 1.2 and Lemma 1.1 we presented two one-square theorems about the triangular numbers $t_n = 1 + 2 + \cdots + n$: $t_{n-1} + t_n = n^2$ and $8t_n + 1 = (2n + 1)^2$. The next theorem presents a two-square theorem for the triangular numbers.

Theorem 8.6. *If N is the sum of two triangular numbers, then $4N + 1$ and $2(4N + 1)$ are each the sum of two squares.*

Proof. Let t_a and t_b be two triangular numbers, $a \geq b \geq 0$, and set $N = t_a + t_b$. Since $t_a = a(a + 1)/2$ and $t_b = b(b + 1)/2$, algebra yields

$$4N + 1 = 4(t_a + t_b) + 1 = (a + b + 1)^2 + (a - b)^2.$$

See Figure 8.10 for an illustration of the case where $(a, b) = (7, 3)$.

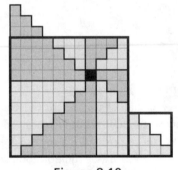

Figure 8.10.

For $2(4N + 1)$, we use Lemma 1.1 to obtain

$$2(4N + 1) = (8t_a + 1) + (8t_b + 1) = (2a + 1)^2 + (2b + 1)^2. \quad ∎$$

The last result (twice a sum of squares is a sum of squares) is actually a special case of Diophantus of Alexandria's sum of squares theorem: *If m and n are each a sum of integral squares, then so is the product mn:*

$$(a^2 + b^2)(c^2 + d^2) = (ac + bd)^2 + (ad - bc)^2,$$

which can be readily verified by expanding each side.

As an application of Theorem 8.6, we prove the following characterization of the integers that can be represented as a sum of two squares [Sutcliffe, 1963].

Theorem 8.7. *An integer M is the sum of two squares $A^2 + B^2$, not both 0, if and only if it is of the form $2^n(4N + 1)$ where N is the sum of two triangular numbers. If A and B have opposite parity, then $n = 0$, if A and B are both odd, $n = 1$, and if A and B are both even, $n > 1$.*

Proof. Let $N = t_a + t_b$, a and b not both zero. Then as a consequence of Theorem 8.6 we have

$$2^{2k}(4N + 1) = 2^{2k}(a + b + 1)^2 + 2^{2k}(a - b)^2$$

and

$$2^{2k+1}(4N + 1) = 2^{2k}(2a + 1)^2 + 2^{2k}(2a + 1)^2,$$

so that $2^n(4N + 1)$ is the sum of two squares with the stated restrictions on n (observe that $a + b + 1$ and $a - b$ have opposite parity).

For the converse, we consider cases.

1. If $A = 2a + 1$ and $B = 2b + 1$, then

$$A^2 + B^2 = (2a + 1)^2 + (2b + 1)^2$$
$$= 2[(a + b + 1)^2 + (a - b)^2].$$

2. If $A = 2^p(2a + 1)$ and $B = 2^q(2b + 1)$ where $q \geq p \geq 1$, then when $q > p$,

$$A^2 + B^2 = 2^{2p}[(2a + 1)^2 + 2^{2(q-p)}(2b + 1)^2],$$

and when $p = q$, we have (using the result in case 1)

$$A^2 + B^2 = 2^{2p+1}[(a + b + 1)^2 + (a - b)^2].$$

So in general the sum of two squares can be reduced to the remaining case, the sum of an even and an odd square, multiplied by 2^n where n satisfies the stated conditions. To complete the proof, we need only show that if

$$(2a)^2 + (2b + 1)^2 = 4N + 1,$$

then N is the sum of two triangular numbers:

$$N = a^2 + b^2 + b$$
$$= \frac{(a + b)(a + b + 1)}{2} + \frac{(a - b - 1)(a - b)}{2}$$
$$= \frac{(a + b)(a + b + 1)}{2} + \frac{(b - a)(b - a + 1)}{2},$$

which expresses N as the sum $t_{a+b} + t_{a-b-1}$ (if $a > b$) or $t_{a+b} + t_{b-a}$ (if $a \leq b$). ∎

When can an integer M be represented as the *difference* of two squares? See Challenge 8.3.

As our final example of a two-square theorem, we present another proof that $\sqrt{2}$ is irrational. This proof was presented at Darwin College, Cambridge, at a lecture in 2002 by John H. Conway of Princeton, who attributes the proof to Stanley Tennenbaum.

As is Section 2.1 assume $\sqrt{2}$ is rational and write $\sqrt{2} = m/n$ where m and n are integers and the fraction is in lowest terms. Then $m^2 = 2n^2$, so that there exist two squares with integer sides such that one has exactly twice the area of the other. See Figure 8.11a.

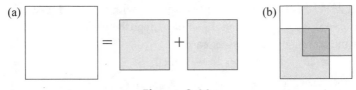

Figure 8.11.

We now place the two smaller gray squares on the larger square, as illustrated in Figure 8.11b. The side of the dark gray square in the center is an integer, and its area must equal the sum of the areas of the two small white squares, whose sides are also integers. Thus we have a contradiction that m and n are the *smallest* integers such that $m^2 = 2n^2$.

8.3   Three-square theorems

As we noted earlier, the Pythagorean theorem is a three-square theorem. Right triangles with integer sides, such as $(a, b, c) = (3, 4, 5)$ or $(5, 12, 13)$ are of particular interest. A triple (a, b, c) of integers such that $a^2 + b^2 = c^2$ is called a *Pythagorean triple* and is called *primitive* whenever a, b, and c have no common factors (such as the two examples above). The following theorem characterizes Pythagorean triples [Teigen and Hadwin, 1971; Gomez, 2005]. In the proof, we superimpose squares with areas a^2 and b^2 on a square with area c^2.

Theorem 8.8. *There is a one-to-one correspondence between Pythagorean triples and factorizations of even squares of the form $n^2 = 2km$.*

Proof. Let (a, b, c) be a Pythagorean triple, and set $k = c - b$, $m = c - a$, and $n = a + b - c$. Then $a = n + k$, $b = n + m$, and $c = n + k + m$, as illustrated in Figure 8.12a. When $a^2 + b^2 = c^2$, the area of the dark gray square in Figure 8.12b equals the sum of the areas of the white rectangles, thus $n^2 = 2km$ and conversely.

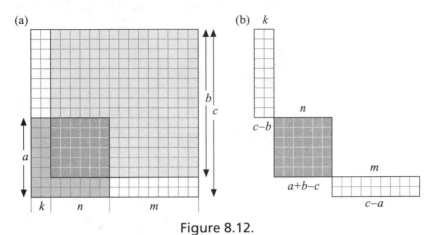

Figure 8.12.

Since $m - k = b - a$, it follows that $a < b$ if and only if $k < m$, and it is easy to show [Teigen and Hadwin, 1971] that (a, b, c) is primitive if and only if k and m are relatively prime. For example, $6^2 = 2 \cdot 1 \cdot 18$ corresponds to the triple (7,24,25), $6^2 = 2 \cdot 2 \cdot 9$ corresponds to (8,15,17), and $6^2 = 2 \cdot 3 \cdot 6$ corresponds to (9,12,15). ∎

Fibonacci and the *Liber Quadratorum*

Leonardo Pisano (1170–1250), or Leonardo of Pisa, is better known by his nickname, Fibonacci. He wrote a several important mathematical texts, the last of which (1225) is the *Liber Quadratorum* (*The Book of Square Numbers*). Although not his best-known work, it is considered by many to be the most impressive. It is a number theory text, and includes, among other things, methods to find Pythagorean triples. Fibonacci's knowledge of Euclid's *Elements* enabled him to solve algebraic problems geometrically, as algebraic symbolism was unknown in his day.

There are several nice results related to the Pythagorean theorem obtained by constructing squares on the sides of a general triangle. Here is one example.

Theorem 8.9. *Given any triangle, construct squares on the sides, and connect the outermost vertices to form three more triangles, as illustrated in Figure 8.13. Then each of the three new triangles has the same area as the original triangle.*

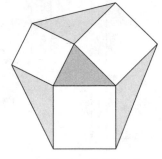

Figure 8.13.

Proof [Snover, 2000]. After erasing the squares, rotate each outer triangle 90° clockwise (see Figure 8.14); and note that each of the rotated triangles has a base and altitude equal to that of the original triangle. ■

(a)　　　　　　　　　　　　　　　　(b)

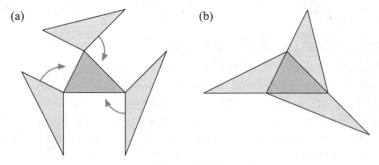

Figure 8.14.

Another three-square theorem using the construction in Figure 8.13 and proved via rotation is the following from [Coxeter and Greitzer, 1967].

Theorem 8.10. *If squares with centers O_1, O_2, O_3 are drawn externally on the sides BC, CA, AB of triangle ABC, then the line segments O_1O_2 and CO_3 are perpendicular and equal in length.* See Figure 8.15.

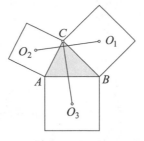

Figure 8.15.

Proof. Draw triangles ABK and CBK as illustrated in Figure 8.16a, and reduce each one in size by a factor of $\sqrt{2}/2$, as illustrated in Figure 8.16b. Note that the images of the segment BK are parallel and equal in length. Now rotate the light gray triangle $45°$ clockwise to ACO_3 and the dark gray triangle $45°$ counterclockwise to CO_1O_2, as shown in Figure 8.16c. As a result, the segments O_1O_2 and CO_3 are perpendicular and equal in length.

∎

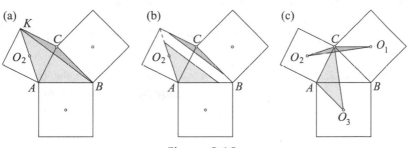

Figure 8.16.

8.4   Four and more squares

The extension of the previous theorem to a quadrilateral with four squares on its sides is a lovely and unexpected result known as *van Aubel's theorem*. Our proof of this four-square theorem uses the two-square Finsler-Hadwiger theorem 8.5 twice.

Van Aubel's Theorem 8.11. *If squares are constructed outwardly on the sides of a convex quadrilateral, then the line segments joining the centers of opposite pairs of squares are perpendicular and equal in length.*

Proof. We first apply Theorem 8.5 to the squares in Figure 8.17a. Letting M denote the midpoint of AB, we have $|PM| = |QM|$ and $PM \perp QM$

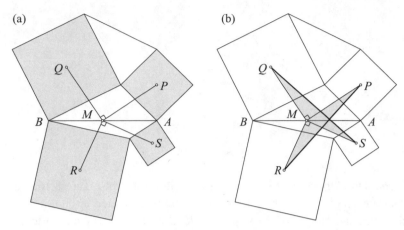

Figure 8.17.

and similarly $|RM| = |SM|$ and $RM \perp SM$. Consequently the gray triangles PMR and QMS are congruent, with corresponding sides perpendicular. Thus $|PR| = |QS|$ and $PR \perp QS$, as claimed. ∎

When two adjacent vertices of the quadrilateral coincide, van Aubel's theorem reduces to Theorem 8.10.

Theorem 8.12. *If two chords of a circle intersect at right angles, then the sum of the areas of the squares whose sides are the four segments formed is constant and equal to the area of the square circumscribed about the circle* (See Figure 8.18).

Proof. Chose axes parallel to the chords, with the origin at the center of the circle. Let the circle be given by $x^2 + y^2 = r^2$, and let the point of

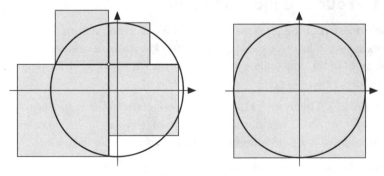

Figure 8.18.

intersection of the two chords be (a, b), where $a^2 + b^2 \le r^2$. The lengths of the four segments are $\sqrt{r^2 - a^2} - b$, $b + \sqrt{r^2 - a^2}$, $\sqrt{r^2 - b^2} - a$, and $a + \sqrt{r^2 - b^2}$ and the sum of the areas of the four squares is (after some simplification)

$$2(r^2 - a^2) + 2b^2 + 2(r^2 - b^2) + 2a^2 = (2r)^2. \qquad \blacksquare$$

It is trivial to partition a given square into four squares—simply draw lines parallel to and midway between the sides. It is also easy to see that it is impossible to partition a square into two, three, or five squares. However, it is amusing to note that all other values are possible, and we show in the next theorem.

Theorem 8.13. *For any $n \ge 6$ a square can be partitioned into n squares.*

Proof. Our proof consists of two parts: (a) first we show it true for $n = 6, 7, 8$; and then (b) show that if it is true for $n = k$, it must be true for $n = k + 3$. See Figure 8.19. $\blacksquare$

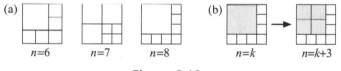

$n=6 \qquad n=7 \qquad n=8 \qquad\qquad n=k \qquad n=k+3$

Figure 8.19.

Other theorems about one, two, three, and four squares (in the geometrical sense) can be found in [DeTemple and Harold, 1996].

8.5 Squares in recreational mathematics

Sam Loyd (1841–1911) was perhaps the best-known creator of mathematical puzzles of his time. The book *Sam Loyd's Cyclopedia of 5000 Puzzles, Tricks, and Conundrums (With Answers)* [Loyd, 1914] contains about 2700 mathematical puzzles and recreations.

Many of Loyd's puzzles are dissection puzzles, where one must dissect one geometrical figure and reassemble the pieces to form a different figure, preferably with as few pieces as possible. Several such puzzles concern the *Greek cross*, a cross in the shape of the plus sign. On page 58 of the *Cyclopedia* Loyd asks us to dissect a square and reassemble the pieces to form a Greek cross (or vice-versa). A five-piece solution can be obtained from the grid in Figure 8.7, it is shown in Figure 8.20a. Loyd always sought to find solutions with as few pieces as possible, and gave the four-piece solution in Figure 8.20b.

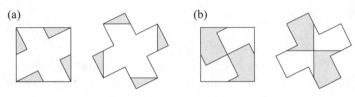

Figure 8.20.

Loyd also considered the dissection of a square into two congruent Greek crosses, and gave the five-piece and four-piece solutions in Figure 8.21.

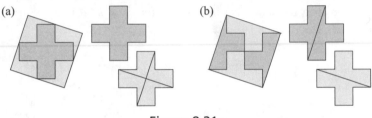

Figure 8.21.

For insight into one method to create dissection puzzles such as Loyd's Greek cross, see Challenge 10.4.

The dissection proofs of the Pythagorean theorem in Figure 5.2 are dissection puzzles, asking for dissections of a square into two squares. It is also possible to dissect a square into three squares, two squares into five squares, and so on. An excellent resource for such recreational problems is [Frederickson, 1997].

Sam Loyd also had problems with multiple squares. A classic example is "The Lake Puzzle," from page 267 of the *Cyclopedia*, which concerns a triangular lake surrounded by three square plots of land (see Figure 8.22). In his own words, Loyd writes: "The question which I ask our puzzlists who revel in just such questions, is to determine just how many acres there would be in that triangular lake, surrounded as shown by three square plots of 370, 116, and 74 acres."

Figure 8.22.

We leave the solution of the Lake Puzzle to Challenge 8.6.

8.6 Challenges

8.1 Verify that the rectangles in Figure 8.4 are 1-by-$1/\sqrt{n}$ root-rectangles.

8.2 Trisect the sides of a square and create a square within a square by drawing a line from one of the trisection points on each side to one of the opposite vertices, as illustrated in Figure 8.23. How does the area of the small gray square compare to the area of the original square in each case?

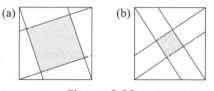

Figure 8.23.

8.3 Show that an integer M can be represented as a difference of integral squares if and only if M is odd or a multiple of 4.

8.4 Show that if $2M$ is a sum of two integral squares, then so is M.

8.5 Use Theorem 8.8 to give another proof that $\sqrt{2}$ is irrational.

8.6 Solve Sam Loyd's Lake Puzzle.

8.7 What is the ratio of the areas of the squares inscribed and circumscribed about the same circle? See Figure 8.24.

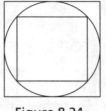

Figure 8.24.

8.8 We have $3(1^2 + 2^2 + 3^2) = 6^2 + 2^2 + 1^2 + 1^2$ and $3(1^2 + 3^2 + 7^2) = 11^2 + 6^2 + 4^2 + 2^2$. Is it true that three times the sum of three squares is always a sum of four squares? [Carroll, 1958]

8.9 Two squares share a vertex, and two of the vertices of the smaller square lie inside the larger, as illustrated in Figure 8.25. Prove that $y = x\sqrt{2}$.

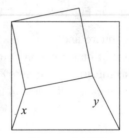

Figure 8.25.

8.10 A point P is located within a square so that its distance from three consecutive vertices of the square is 1, 2, and 3 units, as illustrated in Figure 8.26. What is the degree measure of the angle at P between the segments of lengths 1 and 2?

Figure 8.26.

CHAPTER 9

Curves Ahead

Old Euclid drew a circle
On a sand-beach long ago.
He bounded and enclosed it
With angles thus and so.
His set of solemn greybeards
Nodded and argued much
Of arc and circumference,
Diameter and such.
A silent child stood by them
From morning until noon
Because they drew such charming
Round pictures of the moon.

Vachel Lindsay, *Euclid*

Many mathematical curves have intriguing properties. In visiting the world of curves, one enjoys three complementary views: some curves arise as geometrical shapes appearing in nature, others come from the observation of dynamic phenomena, and a wide range of curves result from mathematical ingenuity [Wells, 1991].

Our aim in this chapter is to present a selection of attractive proofs related to various extraordinary properties of some curves. We begin with some theorems about lunes, a shape that appears in nature as phases of the moon.

9.1 Squarable lunes

As we noted in the previous chapter, the ancient Greeks were concerned with the notion of *quadrature*, constructing with straightedge and compass a square equal in area to a given figure. This process is also called *squaring* the figure, and a figure that can be squared is called *squarable*.

A *lune* (from the French word for moon) is a concave region in the plane bounded by two circular arcs. A convex region bounded by two circular arcs is a *lens*. See Figure 9.1 for an illustration of two lunes (gray) and a lens (white).

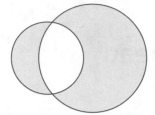

Figure 9.1.

Hippocrates of Chios (c. 470–410 BCE) is believed to be the first person to square lunes. His success in doing so gave hope to squaring the circle, one of the three great geometric problems of antiquity.

Although Hippocrates lived before the time of Euclid, he knew many of the theorems that would later appear in the *Elements* of Euclid. For instance, he knew the generalization of the Pythagorean theorem given in Proposition 31 of Book VI: *In right-angled triangles the figure on the side opposite the right angle equals the sum of the similar and similarly described figures on the sides containing the right angle.* When the figures on the three sides are squares, we have the Pythagorean theorem; when the figures are semicircles, we have the situation illustrated in Figure 6.1b. This is the tool used by Hippocrates to square lunes. The following theorem and its proof are due to Hippocrates.

Theorem 9.1. *If a square is inscribed in the circle and four semicircles constructed on its sides, then the area of the four lunes equals the area of the square.* See Figure 9.2.

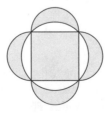

Figure 9.2.

Proof. See Figure 9.3. ∎

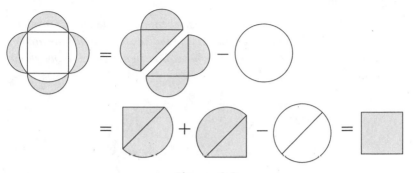

Figure 9.3.

Dividing Figure 9.2 into two congruent halves along a diagonal of the square shows that the area of two lunes on the legs of an isosceles right triangle equals the area of the triangle. Hippocrates proved that the same is true for any right triangle.

Theorem 9.2. *The combined area of the lunes constructed on the legs of a right triangle is equal to the area of the triangle.*

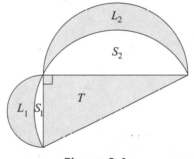

Figure 9.4.

Proof. In Figure 9.4 [Margerum and McDonnell, 1997], let L_1, L_2, T, S_1, S_2 denote the areas of the two lunes, the triangle, and the two circular segments, respectively. The semicircular version of the Pythagorean theorem yields $(L_1 + S_1) + (L_2 + S_2) = (T + S_1 + S_2)$, from which $L_1 + L_2 = T$ follows. ∎

These lunes are also known as the *Lunes of Alhazen*, since they appear in the works of Abu Ali al-Hasan ibn al-Hasan ibn al-Haytham al-Basri (965–1040).

Hippocrates also established the following result relating the areas of a hexagon, six lunes, and a circle.

Theorem 9.3. *If a regular hexagon is inscribed in a circle and six semicircles constructed on its sides, then the area of the hexagon equals the area of the six lunes plus the area of a circle whose diameter is equal to one of the sides of the hexagon.*

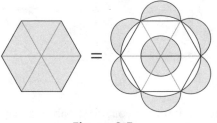

Figure 9.5.

Proof. See Figure 9.6 [Nelsen, 2002c]. In this proof, we use the fact (known to Hippocrates) that the area of a circle is proportional to the square of its radius, so that the four small gray circles in the second line have the same combined area as the large white circle. ■

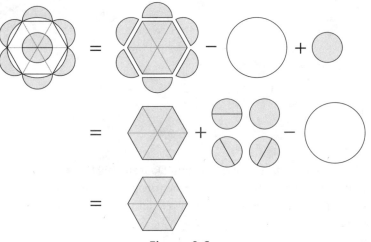

Figure 9.6.

Unlike the lunes in Figures 9.2 and 9.4, the lunes in Figure 9.5 are not squarable. If they were, then by subtraction, the circle in the center of the hexagon would also be squarable, which it is not. However, this was not known to Hippocrates, since the impossibility of squaring the circle was not established rigorously until Carl Louis Ferdinand von Lindemann (1852–1939) did so in 1882 by proving that π is transcendental.

The *lunulae* that fascinated Leonardo da Vinci

Immediately after the November 10, 1494 publication of Luca Pacioli's *Summa de Arithmetica, Geometria, Proportioni et Proportionalità*, Leonardo bought a copy. He was fascinated by the problems of squaring the circle and the *lunulae* (Latin for lunes). He studied the work of Pacioli and did his own research, the results of which appear in the *Madrid Code* (8936) and in the *Atlantic Code* (folio 455 recto).

In 1496 Leonardo and Luca met in Milan, and as a consequence Leonardo became interested in other geometrical problems. However, none matched his interest in and skill with the squaring of curvilinear figures (See Challenge 9.5).

And so, beyond the lunes in Figures 9.2 and 9.4, what other lunes are squarable? To be precise, we are interested in *constructible squarable lunes*, the lunes for which we can construct a square with the same area using only a straightedge and compass. Hippocrates found three (up to similarity), Martin Johan Wallenius found two more in 1766, and during the period between 1934 and 1947, Chebotarov and Dorodnov proved that there are no other squarable lunes. For details, see [Postnikov and Shenitzer, 2000].

The lunes of Hippocrates in Figure 9.4 are the basis for other interesting quadrature results.

Theorem 9.4. *Let T denote the area of the large right triangles in Figure 9.7a and 9.7b. Then we have* (i) $A - B_1 - B_2 = T$ *in Figure 9.7a and* (ii) $A + B + C + D = T$ *in Figure 9.7b* [Gutierrez, 2009].

Proof. Let T_1 denote the area of the right triangle to the left of the dashed altitude to the hypotenuse, and T_2 the area of the right triangle to the right of the dashed altitude, so that $T_1 + T_2 = T$. In (i), $B_2 + u + T_1$ and $B_1 + v + T_2$ are the areas of semicircles on the legs of a right triangle, so they sum to the area $A + u + v$ of a semicircle on the hypotenuse. In (ii), $(A + B) + (C + D) = T_1 + T_2 = T$. ∎

In addition to lunes, other figures bounded by arcs of circles are squarable. For a simple example, consider the shaded region in Figure 9.8a. It has the same area as the rectangle in Figure 9.8b.

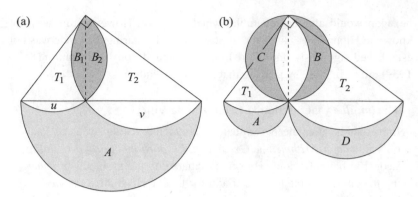

Figure 9.7.

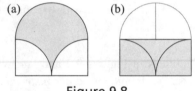

Figure 9.8.

The same principle extends to other figures constructed from polygons and arcs of circles, as illustrated in Figure 9.9 for a triangle and a square.

These figures underlie some of the remarkable tilings found in Moorish and Mudéjar palaces in Andalucia, Spain, and in some of the Moorish-inspired work of the architect Antoni Gaudí. The left hand tiling in Figure 9.10 is from Gaudí's Pavellons Güell in Barcelona, Spain, and the other two are from walls in the Real Alcázar in Seville. In each case the tile is a squarable figure.

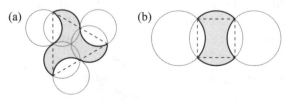

Figure 9.9.

Figure 9.10.

Similar to lunes are the crescents that appear in certain puzzles created by Sam Loyd (1841–1911) and Henry Ernest Dudeney (1857–1930). These dissection puzzles ask us to cut a crescent into several pieces that can be reassembled to form a Greek cross (see Section 8.5). In these puzzles, a *crescent* is a region bounded by two congruent circular arcs and two parallel line segments of equal length. In Figure 9.11 we see Loyd's "Cross and Crescent" puzzle in which the distance between the line segments is five times their length.

Figure 9.11.

One method for solving the cross and crescent puzzle is to use a tiling with crescents and an overlay tiling with crosses, as illustrated in Figure 9.12. For details about this dissection, see [Frederickson, 1997].

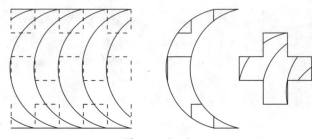

Figure 9.12.

9.2 The amazing Archimedean spiral

Archimedes introduced the spiral bearing his name in about 225 BCE for the express purpose of squaring the circle and trisecting angles (contrary to popular belief, several Greek geometers used tools other than the straightedge and compass for these problems). In dynamical terms, the Archimedean spiral is the locus of a point traveling on a ray at a uniform rate while (at the same time) the ray rotates at a uniform rate. The polar equation for the spiral is $r = a\theta$, i.e., the distance of the point from the origin is proportional to the angle of rotation. See Figure 9.13.

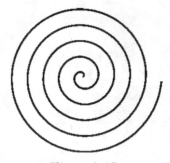

Figure 9.13.

Figure 9.13 shows just one branch of the spiral, the branch for $\theta \geq 0$. The other branch ($\theta \leq 0$) is the refection of the $\theta \geq 0$ branch in the $\pi/2$- (or y-) axis. In what follows we need only consider the $\theta \geq 0$ branch of the spiral. The r-coordinates of the points of intersection of the spiral with any ray from the origin form an arithmetic progression with common difference 2π.

We now show that how the Archimedean spiral can be used to square circles and multisect angles.

Theorem 9.5. *The Archimedean spiral squares circles.*

Proof. Given the circle $r = a$ of radius a centered at the origin, it suffices to construct a rectangle with area πa^2. Draw the Archimedean spiral $r = a\theta$, as shown in Figure 9.14. The spiral intersects the $\pi/2$-axis at a point $a\pi/2$ units from the origin, so that we can construct the shaded rectangle with height $a\pi/2$, base $2a$, and area πa^2 as required. ∎

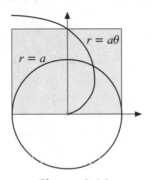

Figure 9.14.

Theorem 9.6. *The Archimedean spiral multisects angles.*

Proof [Aczél and Alsina, 1998]. We will show how the Archimedean spiral trisects angles; the proof for multisection is analogous. Let $\angle AOB$ be the angle to be trisected, and place O at the origin and OA on the polar axis, as illustrated in Figure 9.15. Let α denote the measure of $\angle AOB$. Draw the Archimedean spiral $r = \theta$. The spiral will intersect OA at $P = (\alpha, \alpha)$. With

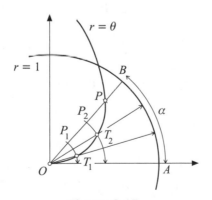

Figure 9.15.

a straightedge and compass divide the segment OP into 3 parts, and let P_i be the point on OP such that $|OP_i| = i\alpha/3, i = 1, 2$. Now rotate each segment OP_i until it cuts the spiral at the point $T_i = (i\alpha/3, i\alpha/3)$. Drawing the line segments $OT_i, i = 1, 2$ completes the trisection. ∎

The ability to multisect angles enables us to inscribe n-gons in circles for any $n \geq 3$—we simply use the Archimedean spiral to construct n rays from the center to the circle, and join the endpoints to form the n-gon.

> **Archimedean spirals today**
>
> Archimedean spirals occur (approximately) in many man-made objects: rolls of paper and garden hoses, coiled watch balance springs, and the grooves on old-fashioned LP vinyl phonograph records. They also appear in mechanical devices such as the scroll compressors used to compress gases in air-conditioning equipment and automobile super-chargers.

9.3   The quadratrix of Hippias

Perhaps the first curve described in mathematics after the line and the circle was the *trisectrix* of Hippias of Elis (circa 460–400 BCE) for the purpose of trisecting angles. Later Pappus of Alexandria (circa 290–350 CE) demonstrated that the trisectrix could also be use to square circles, hence the trisectrix became known as the *quadratrix* of Hippias.

The quadratrix is most easily described dynamically. Start with a square of area 1 as shown in Figure 9.16a. Now let OA rotate clockwise about O at a uniform rate (OA' in the figure) at the same time as AB descends at a uniform rate ($A''B'$ in the figure), so that both OA' and $A''B'$ coincide with

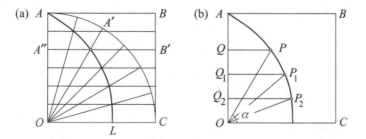

Figure 9.16.

OC simultaneously. The locus of the intersection of OA' and $A''B'$ is the quadratrix.

It is now a simple matter to use the quadratrix of Hippias to trisect angles; the procedure for multisection is similar. It suffices to trisect angles with measure $\alpha \in (0, \pi/2)$. Let $\angle POC = \alpha$ as shown in Figure 9.16b, and draw QP parallel to OC. Now trisect OQ to locate points Q_1 and Q_2, and draw Q_1P_1 and Q_1P_1 parallel to OC. Then, by the dynamic definition of the quadratrix, $\angle POP_1$, $\angle P_1OP_2$, and $\angle P_2OC$ each have measure $\alpha/3$.

It is easy to show that a Cartesian equation of the quadratrix of Hippias is $x = y \cot(\pi y/2)$ for $y \in (0, 1]$ and $x = 2/\pi$ for $y = 0$ (see Challenge 9.7). As a consequence, we can now square circles! We shall only square the unit circle; others can be squared similarly. Since $|OL| = 2/\pi$, we can construct a segment of length $\pi/2$ using similar triangles, and then a rectangle (and thus a square) of area π.

9.4 The shoemaker's knife and the salt cellar

In Greek, a shoemaker's knife is an *arbelos* ($\alpha\rho\beta\epsilon\lambda os$), and a salt cellar is a *salinon* ($\sigma\alpha\lambda\iota\nu o\nu$). Archimedes used these words to describe two plane figures bounded by semicircles, and also described how to easily compute their areas.

The arbelos appears in Proposition 4 of his *Liber Assumptorum* (*Book of Lemmas*), where Archimedes writes:

Theorem 9.7. *Let P, Q, and R be three points on a line, with Q lying between P and R. Semicircles are drawn on the same side of the line with diameters PQ, QR, and PR. An arbelos is the figure bounded by these three semicircles. Draw the perpendicular to PR at Q, meeting the largest semicircle as S. Then the area A of the arbelos equals the area C of the circle with diameter QS. See Figure 9.17.*

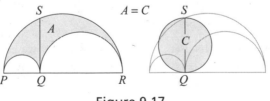

Figure 9.17.

Proof [Nelsen, 2002b]. Using the semicircular version of the Pythagorean theorem (see Sections 6.1 and 9.1), we have, in Figure 9.18a, $A + A_1 + A_2 = B_1 + B_2$, in Figure 9.18b, $B_1 = A_1 + C_1$, and in Figure 9.18c, $B_2 = A_2 + C_2$.

Together the equations yield $A + A_1 + A_2 = A_1 + C_1 + A_2 + C_2$, which simplifies to $A = C_1 + C_2 = C$. ∎

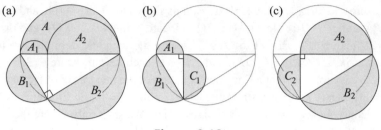

Figure 9.18.

The salinon appears in Proposition 14 of the *Liber Assumptorum*:

Theorem 9.8. *Let P, Q, R, S be four points on a line (in that order) such that* $PQ = RS$. *Semicircles are drawn above the line with diameters PQ, RS, and PS, and another semicircle with diameter QR is drawn below the line. A* salinon *is the figure bounded by these four semicircles. Let the axis of symmetry of the salinon intersect its boundary at M and N. Then the area A of the salinon equals the area C of the circle with diameter MN. See Figure 9.19.*

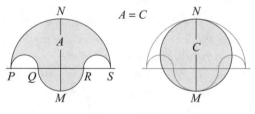

Figure 9.19.

Proof [Nelsen, 2002a]. [Nelsen, 2002a]. In this proof, we employ the fact that the area of a semicircle is $\pi/2$ times the area of the inscribed isosceles right triangle (see Figure 9.20).

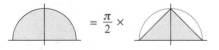

Figure 9.20.

Consequently, the area of the salinon is $\pi/2$ times the area of a square (see Figure 9.21), which equals the area of a circle and proves the theorem. ∎

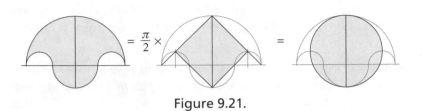

Figure 9.21.

Neither the arbelos nor the salinon is a squarable figure, since each is equivalent in area to a circle.

9.5 The Quetelet and Dandelin approach to conics

Parabolas, ellipses, and hyperbolas are remarkable curves originally studied by the Greeks as plane sections of a cone. In analytic geometry we learn that the graph of an equation of the second degree, $ax^2 + bxy + cy^2 + dx + ey + f = 0$, is either one of the three conics, a line, a pair of lines, a point, or imaginary. Conics play an important role in both pure and applied mathematics, for example as models for orbits of objects from electrons to planets.

In Figure 9.22 we see both nappes of three cones, and a parabola, an ellipse, a circle, and a hyperbola as conic sections.

Figure 9.22.

It is common in courses in analytic geometry and calculus to study the conics using the focus-directrix property rather the conic section method of the Greeks. The following theorem and elegant proof of the equivalence of the two approaches is due to Adolphe Quetelet (1796–1874) and Germinal-Pierre Dandelin (1794–1847) [Eves, 1983]. The following lemma is essential to our proof.

Lemma 9.1. *The lengths of any two line segments from a point to a plane are inversely proportional to the sines of the angles that the line segments make with the plane.*

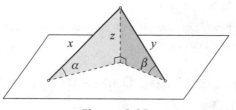

Figure 9.23.

Proof. See Figure 9.23, and observe that $z = x \sin \alpha = y \sin \beta$, hence $x/y = \sin \beta / \sin \alpha$. ∎

Theorem 9.9. *Let π denote a plane that intersects a right circular cone in a conic section, and consider a sphere tangent to the cone and tangent to π at a point F (see Figure 9.24). Let π' denote the plane determined by the*

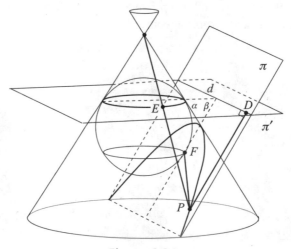

Figure 9.24.

circle of tangency of the sphere and the cone, and let d denote the line of intersection of π *and* π'. *Let P be any point on the conic section, and let D denote the foot of the line segment from P perpendicular to d. Then the ratio* $|PF|/|PD|$ *is a constant.*

Proof. Let E denote the point of intersection of the element of the cone (a line on the cone passing through the vertex) through P and the circle of tangency of the sphere. Then $|PF| = |PE|$. Let α denote the angle that every element of the cone makes with π', and let β denote the angle between π and π'. Then $|PF|/|PD| = |PE|/|PD| = \sin\beta/\sin\alpha$, and $\sin\beta/\sin\alpha$ is a constant. ∎

The point F in the proof is the *focus* of the conic section, and the line d the *directrix*. The constant $\sin\beta/\sin\alpha$ is often denoted by e, the *eccentricity* of the conic section. When π is parallel to one and only one element of the cone, $\alpha = \beta$, $e = 1$, and the conic is a parabola; when π cuts every element of one nappe of the cone, $\alpha > \beta$, $e < 1$, and the conic is an ellipse; when π cuts both nappes of the cone, $\alpha < \beta$, $e > 1$, and the conic is a hyperbola.

9.6 Archimedes triangles

Among the great discoveries of Archimedes was the *quadrature of the parabola*, that is, a parabolic section such as the shaded region in Figure 9.25a is squarable. His tool for accomplishing this was a special kind of triangle now known as an *Archimedes triangle*, one whose sides are tangents to a parabola and the chord joining the points of tangency (SA, SB, and AB, respectively, in Figure 9.25a). We shall refer to the tangents SA and SB as the sides and the chord AB as the base of the Archimedes triangle. Archimedes then proved (as we shall) that the area of the shaded parabolic segment in Figure 9.25a is 2/3 of the area of triangle SAB.

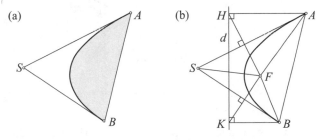

Figure 9.25.

How does one construct an Archimedes triangle? The procedure is illustrated in Figure 9.25b for points A and B on a given parabola with focus F and directrix d. Draw AH and BK perpendicular to d, then draw FH and FK. Since $|FA| = |AH|$ and $|FB| = |BK|$, triangles FAH and FBK are isosceles. Now draw the perpendicular bisectors of FH and FK, extended to meet in the point S. SA and SB also bisect the apex angles FAH and FBK of the two isosceles triangles and hence are tangent to the parabola. Drawing the chord AB completes our construction of the Archimedes triangle SAB.

We now prove a lemma that will enable us to square parabolic segments.

Lemma 9.2. *In an Archimedes triangle,* (i) *the median to the base is parallel to the axis, and* (ii) *the line tangent to the parabola at the point where the median intersects the parabola bisects the other two sides of the triangle and is parallel to the base.*

Proof. Refer to Figure 9.25b. Since two of the perpendicular bisectors of the sides of the triangle FHK meet at S, so does the perpendicular bisector of the third side HK. This bisector is parallel to AH and BK, and hence to the axis of the parabola, and it also bisects AB, which proves (i). In Figure 9.26a let O be the point of intersection of the median SM to the base with the parabola, and draw $A'B'$ through O tangent to the parabola. Thus $AA'O$ and $BB'O$ are Archimedes triangles, and so the medians to their bases (dashed segments in Figure 9.26a) are parallel to the axis and hence to SM. In triangle SAO the dashed segment bisects AO and is parallel to SO, and hence A' is the midpoint of SA. Similarly B' is the midpoint of SB, thus $A'B'$ is parallel to AB, which proves (ii). ∎

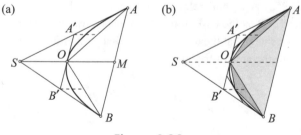

Figure 9.26.

Theorem 9.10. *The area of a parabolic section is two-thirds the area of the corresponding Archimedes triangle.*

Proof. Let K denote the area of the Archimedes triangle SAB. Then the area of triangle AOB (light gray in Figure 9.26b) inscribed in the parabolic section is $K/2$, the area of triangle $SA'B'$ is $K/4$, and hence the triangles

$AA'O$ and $BB'O$ each have area $K/8$. Thus the two dark gray triangles inscribed in the parabolic section each have area $(1/8)(K/2)$. The next iteration will yield four inscribed triangles each with area $(1/8^2)(K/2)$. Continuing in this fashion and summing yields the infinite series for the area of the parabolic section:

$$\frac{K}{2} + 2 \cdot \frac{1}{8} \cdot \frac{K}{2} + 4 \cdot \frac{1}{8^2} \cdot \frac{K}{2} + \cdots = \frac{K}{2}\left(1 + \frac{1}{4} + \frac{1}{4^2} + \cdots\right) = \frac{K}{2} \cdot \frac{4}{3} = \frac{2K}{3}.$$

∎

Any given triangle is an Archimedes triangle for three different parabolas, as illustrated in Figure 9.27a. This configuration has many lovely properties; we mention only two in the next theorem.

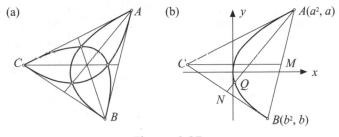

Figure 9.27.

Theorem 9.11. *Consider an Archimedes triangle for three parabolas, each parabola tangent to two sides of the triangle at vertices. Then* (i) *the parabolas always intersect on medians of the triangle, and* (ii) *the points of intersection divide each median into two segments in the ratio* 8:1.

Proof [Bullard, 1935; Honsberger, 1978]. See Figure 9.27b. We prove both parts of the theorem together by showing that the point Q that divides the median AN into two segments in the ratio 8:1 lies on the parabola shown. Introduce an axis system with origin at the vertex and the x-axis the axis of the parabola, and chose a scale so that the equation of the parabola is $x = y^2$. Let A and B be (a^2, a) and (b^2, b), respectively. Then the y-coordinate of M is $(a + b)/2$, and since the equation of AC is $x - 2ay + a^2 = 0$, the coordinates of C are $(ab, (a + b)/2)$. Since N is the midpoint of BC, its coordinates are $(b(a + b)/2, (a + 3b)/4)$. Since Q divides AN into the ratio 8:1, its x-coordinate is

$$\frac{1}{9}a^2 + \frac{8}{9} \cdot \frac{b(a + b)}{2} = \frac{1}{9}(a^2 + 4ab + 4b^2) = \left(\frac{a + 2b}{3}\right)^2,$$

its y-coordinate is

$$\frac{1}{9}a + \frac{8}{9} \cdot \frac{a+3b}{4} = \frac{1}{9}(3a+6b) = \frac{a+2b}{3},$$

and thus Q lies on the parabola $x = y^2$. ∎

For additional properties of Archimedes triangles for three parabolas, see [Bullard, 1935, 1937].

9.7 Helices

A *circular helix* (or coil) is a curve that lies on the surface of a right circular cylinder with the property that tangents to the curve intersect the elements of the cylinder at a constant angle. Its parametric equations are $x = r\cos\theta$, $y = r\sin\theta$, $z = k\theta$ for θ in $[0, 2T\pi]$, where θ is the parameter and r, k, and T are constants: r is the radius of the cylinder, $2\pi k$ the vertical separation between the loops of the helix, and T the number of revolutions around the cylinder. In Figure 9.28 we see a helix with $r = 1$, $k = 1$, and $T = 2$.

The projection of a circular helix onto a plane parallel to its axis (the axis of the cylinder) is a sinusoidal curve.

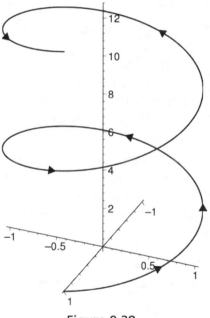

Figure 9.28.

Theorem 9.12. *The shortest path between two points on a cylinder is either a straight line or a fractional turn of a helix.*

Proof. If the two points lie on the same element of the cylinder, then the shortest path is the line segment joining them. Otherwise we can cut the cylinder along an element not passing through either point, and lay it out flat, as shown in Figure 9.29. The shortest path between two points in the rectangle is a portion (i.e., a fractional turn) of some helix on the cylinder. A curious fact is that if one cuts the cylinder along the helix (rather than on an element of the cylinder) and lays it out flat, the resulting plane figure is a parallelogram. ■

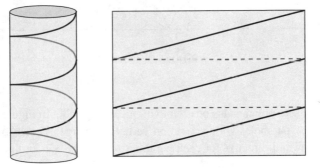

Figure 9.29.

Figure 9.29 also aids us in finding the length of a circular helix. The length of the helix once around the cylinder is the length of the hypotenuse of a right triangle with legs $2\pi r$ and $2\pi k$, that is, $2\pi \sqrt{r^2 + k^2}$, thus the total length for given values of r, k, and T is $2\pi T \sqrt{r^2 + k^2}$.

Helices everywhere

Helices abound in nature—from the structure of the DNA molecule to the shapes of some animal horns to the paths taken by squirrels as they chase one anther around a tree trunk. Helices are also common in man-made objects such as screws, bolts, springs, staircases, and helical antennas.

9.8   Challenges

9.1 In the Cartesian plane let O be the origin, C the circle of radius a centered at $(0,a)$, and T the line tangent to C at $(0,2a)$. The *cissoid of Diocles* is the locus of points P such that $|OP| = |AB|$ where A and B are the points of intersection of OP with C and T, respectively. See Figure 9.30. Find the equation of the cissoid, and show how it can be used to duplicate the cube, that is, to represent the cube root of 2.

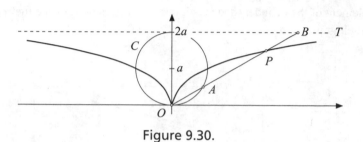

Figure 9.30.

9.2 Conic sections can also be used to duplicate the cube. In Figure 9.31 we see two parabolas with a common vertex and perpendicular axes. Find a focus and directrix for each parabola so that the x-coordinate of the point of intersection is $\sqrt[3]{2}$.

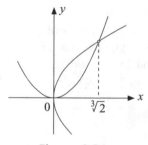

Figure 9.31.

9.3 Show that a lune constructed on the side of an equilateral triangle is not squarable.

9.4 Show that the shaded portions of the objects in Figure 9.32 are squarable (the curves are arcs of circles).

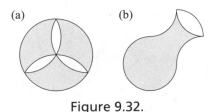

Figure 9.32.

9.5 *Leonardo's claw* is the gray portion of the circle in Figure 9.33a remaining when a smaller circle and a lens, defined by an isosceles right triangle whose legs are radii of the circle, have been removed. Show that Leonardo's claw is squarable, and that its area is the same as the area of the square in its grasp (see Figure 9.33b).

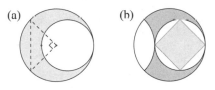

Figure 9.33.

9.6 Prove the *broken chord theorem of Archimedes*: If AB and BC are two chords in a circle with $|AB| > |BC|$ and M is the midpoint of arc ABC, then the foot F of the perpendicular line from M to BC is the midpoint of the broken chord ABC

9.7 Show that a Cartesian equation of the quadratrix of Hippias is given by $x = y \cot(\pi y/2)$ for $y \in (0, 1]$ and $x = 2/\pi$ for $y = 0$.

9.8 Consider a circular disk partitioned into four regions by semicircular arcs, as illustrated in Figure 9.34. The boundaries of the regions partition the horizontal diameter into four intervals of equal width. Prove that the regions have equal area [Esteban, 2004].

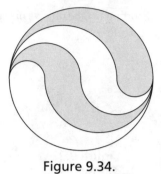

Figure 9.34.

9.9 Let Γ be the graph of a differentiable function concave on the interval $[a, b]$. Prove that the point at which the tangent line to Γ minimizes the shaded area in Figure 9.35 is the midpoint of $[a, b]$. (Hint: no calculus required!)

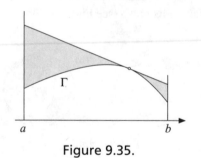

Figure 9.35.

CHAPTER **10**

Adventures in Tiling
and Coloring

> *"What's the color got to do with it?"*
>
> *"It's got everything to do with it. Illinois is green, Indiana is pink.*
>
> *"Indiana PINK? Why, what a lie!"*
>
> *"It ain't no lie; I've seen it on the map, and it's pink."*
>
> Mark Twain
> *Tom Sawyer Abroad*

There are many lovely proofs in which the *tiling* of regions and the *coloring* of objects appear. We have seen examples in earlier chapters, such as tiling in the proof of Napoleon's theorem in Section 6.5, and coloring in the proof of the art gallery theorem in Section 4.6. Tiles enable us to compare areas without calculation, and color enables us to distinguish relevant parts of figures easily.

In this chapter we continue our exploration of elegant proofs that employ these techniques. We begin with a brief survey of basic properties of tilings, followed by some results using tilings with triangles and quadrilaterals, including the Pythagorean theorem. We also discuss the lovely tilings found in the Alhambra in Granada, Spain, and in the work of M. C. Escher. After visiting the seven frieze patterns, we examine the use of color in proofs, and proofs about colorings. Our examples include tiling chessboards with polyominoes, packing calissons into boxes, map coloring, and Hamiltonian circuits on dodecahedra.

10.1 Plane tilings and tessellations

A *tiling* of the plane is any countable family of closed sets (the *tiles*) that cover the plane without gaps or overlaps [Grünbaum and Shephard, 1987]. In this section and ones to follow, our tiles will be polygons. An *edge-to-edge* tiling is a polygonal tiling in which each edge of each polygon coincides with an edge of another polygon. Non-edge-to-edge tilings lead to some unexpected proofs, as we shall see in the next section. A *vertex* of an edge-to-edge tiling is a vertex of one of the tiles. An edge-to-edge tiling in which all the tiles are regular polygons is called a *tessellation*. A *monohedral* tiling is one in which all the tiles are congruent, and a *uniform* tessellation is one with identical vertices, i.e., one with the same arrangement of polygons at each vertex.

We begin by considering *regular tilings*, the monohedral tessellations—edge-to-edge tilings with congruent polygons. Clearly (as Figure 10.1 shows) equilateral triangles, squares, and regular hexagons yield regular tilings.

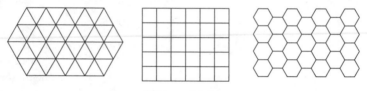

Figure 10.1.

Are there others? The negative answer is given in the next theorem.

Theorem 10.1. *The only regular tilings consist of equilateral triangles, squares, and regular hexagons.*

Proof. Suppose kn-gons meet at a vertex. Since each interior vertex angle of an n–gon measures $(1 - 2/n)\,180°$, we have $k\,(1 - 2/n)\,180° = 360°$, or $(n - 2)(k - 2) = 4$. The only solutions to this equation in positive integers are $(n, k) = (3, 6)$, $(4,4)$, and $(6,3)$. ∎

The regular tilings are clearly uniform tessellations. We can generalize regular tilings in several ways, two of which are (i) use more than one kind of regular polygon (a tessellation, but not monohedral), or (ii) use identical nonregular polygons (monohedral, but not a tessellation). One class of non-monohedral tessellations is the class of uniform non-monohedral tessellations, called *semiregular* (or *Archimedean*) tilings. To find all the semiregular tilings, we need first the following lemma.

Lemma 10.1. *The number of regular polygons that may share a common vertex in an edge-to-edge tiling is 3, 4, 5, or 6.*

Proof. Suppose k polygons share a common vertex in a given edge-to-edge tiling. Clearly $k \geq 3$. Let $\alpha_1, \alpha_2, \ldots, \alpha_k$ denote the vertex angles of the polygons. Then each α_i is at least $60°$ so that $360° = \alpha_1 + \alpha_2 + \cdots + \alpha_k \geq k \cdot 60°$, or $k \leq 6$. ∎

In Figure 10.2 we see eight semiregular tilings. In Theorem 10.2 we prove that there are no others.

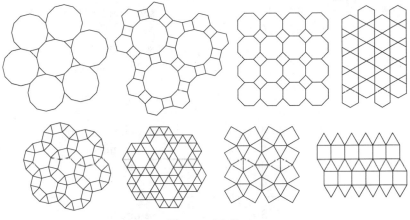

Figure 10.2.

Theorem 10.2. *There are eight classes of semiregular tilings.*

Proof. Since semiregular tilings are uniform, at each vertex we have k regular polygons with $n_1, n_2, \ldots, n_k$ sides, $3 \leq k \leq 6$ and each $n_i \geq 3$. Since the interior vertex angles of the polygons are $(1 - 2/n_i)\,180°$, we have $\sum_{i=1}^{k} (1 - 2/n_i)\,180° = 360°$, or equivalently

$$\sum_{i=1}^{k} \frac{1}{n_i} = \frac{k-2}{2}.$$

The solutions to these four equations ($k = 3$, 4 5, 6) are given in Table 10.1.

The six numeric solutions marked ✗ do not yield tilings, since in a tiling if one of $\{n_1, n_2, n_3\}$ is odd, then the other two must be equal, as those tiles alternate around the tile with an odd number of sides. Also, the solution (3,3,3,4,4) corresponds to two semiregular tilings, one with the two squares adjacent and one where they are not adjacent. ∎

Table 10.1.

k	n_1	n_2	n_3	n_4	n_5	n_6	Conclusion
	3	7	42				✗
	3	8	24				✗
	3	9	18				✗
	3	10	15				✗
3	3	12	12				semiregular (1)
	4	5	20				✗
	4	6	12				semiregular (2)
	4	8	8				semiregular (3)
	5	5	10				✗
	6	6	6				regular
	3	3	4	12			not uniform
4	3	3	6	6			semiregular (4)
	3	4	4	6			semiregular (5)
	4	4	4	4			regular
5	3	3	3	3	6		semiregular (6)
	3	3	3	4	4		semiregular (7 & 8)
6	3	3	3	3	3	3	regular

We now turn our attention to monohedral tilings that are not tessellations. We begin with triangles and quadrilaterals. It is convenient to consider quadrilaterals first.

Theorem 10.3. *Any (convex or concave) quadrilateral tiles the plane.*

Proof. See Figure 10.3. ∎

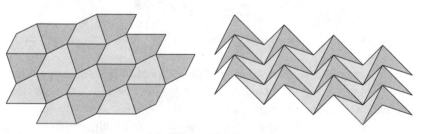

Figure 10.3.

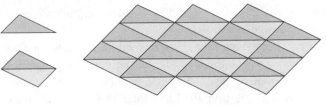

Figure 10.4.

Since any triangle can be doubled to form a parallelogram (see Figure 10.4) and parallelograms are quadrilaterals, the following corollary is immediate

Corollary 10.1. *Any triangle tiles the plane.*

In the next section we explore the use of these tilings in the proofs of some theorems about triangles and quadrilaterals.

What about pentagons? Clearly regular pentagons do not tile the plane (although they do form the surface of a regular dodecahedron in three dimensions), but some non-regular pentagons do. For example, one can partition each hexagon in the regular hexagonal tiling in Figure 10.1 into two or three congruent pentagons to obtain the monohedral pentagonal tilings in Figure 10.5.

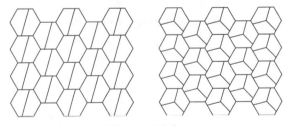

Figure 10.5.

Another pentagonal tiling can be created by overlaying two non-regular hexagonal tilings illustrated in Figure 10.6. This rather attractive monohedral

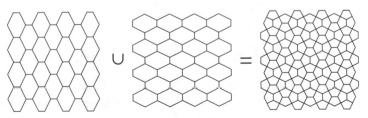

Figure 10.6.

pentagonal tiling is sometimes called the *Cairo tiling*, for its reported use as a street paving design in that city.

Thus we arrive at the question: How many different convex pentagons tile the plane? The answer to this question has an interesting history [Klarner, 1981]. Five different pentagons that tile (including the three above) were described by K. Reinhardt in 1918 and three more by R. B. Kerschner in 1968. A ninth pentagon that tiles was found by R. James in 1975, and four more by Marjorie Rice, an amateur mathematician and homemaker, in 1976–77. A fourteenth pentagon that tiles the plane was found by R. Stein in 1985.

That is where things stand now. Since 1985 no new pentagons that tile the plane have been discovered, nor has a proof appeared that the classification is complete. An answer to this open problem would close the study of mono-hedral tilings, since, as we shall see, the classification for n-gons with $n \geq 6$ is known.

As with pentagons, not all convex hexagons tile the plane. In 1918, K. Reinhardt described the three classes of convex hexagons that tile the plane [Gardner, 1988]. Let A, B, C, D, E, and F denote the vertices, and a, b, c, d, e, and f the sides, as illustrated in Figure 10.7. Then the hexagon tiles the plane if and only if it belongs to one of three classes: I. $A + B + C = 360°$; II. $A + B + D = 360°$, $a = d$, $c = e$; III. $A = C = E = 120°$, $a = b$, $c = d$, $e = f$.

Finally, any convex n-gon with $n \geq 7$ cannot tile the plane [Kerschner, 1969; Niven, 1978], which concludes our brief survey of tilings and tessellations.

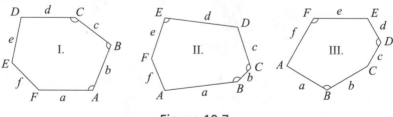

Figure 10.7.

10.2   Tiling with triangles and quadrilaterals

As noted in the previous section, every triangle—acute, right-angled, or obtuse—tiles the plane, as does every quadrilateral—convex or concave.

These facts enable us to use tiling to give simple, visual proofs of some theorems about triangles and quadrilaterals.

Our first result concerns the *median triangle* associated with an arbitrary triangle. It is the triangle constructed from the three medians of the given triangle, as illustrated in Figure 10.8. The next theorem [Hungerbühler, 1999] relates the area of the median triangle to the area of the original triangle.

Theorem 10.4. *The median triangle has* 3/4 *the area of the original triangle.*

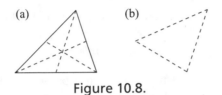

Figure 10.8.

Proof. In Figure 10.9 we have drawn a small portion of the tiling of the plane by the triangle in Figure 10.8a, and overlaid a triangle each of whose sides is twice as long as a median of the given triangle. As a consequence, the medians of a triangle will always form a triangle.

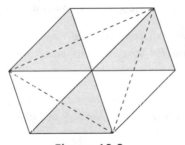

Figure 10.9.

Each median of a triangle partitions the triangle into two smaller triangles, each with half the area of the original triangle. Thus the area of the overlaid triangle is three times the area of the original triangle and four times the area of the median triangle, which proves the theorem. ∎

What if we replace the medians of the triangle with lines (called *cevians*) from the vertices to points on the sides one-third the distance from one vertex to the next? These cevians no longer intersect in a point, but rather form a

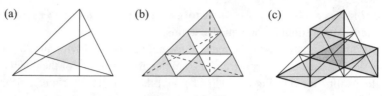

(a)   (b)   (c)

Figure 10.10.

small triangle within the original triangle, as illustrated in Figure 10.10a. How does the area of the small triangle compare with the area of the original triangle?

Theorem 10.5. *If the vertices of a triangle are joined by lines to the one-third points on opposite sides, then the triangle formed in the interior has one-seventh the area of the original triangle.*

Proof [Johnston and Kennedy, 1993]. In this proof, we tile the interior of the given triangle with smaller triangles generated by the trisections of the side, then overlay a grid formed by lines parallel to the line drawn from the vertices to the one-third points. See Figures 10.10b and 10.10c. The seven gray triangles in Figure 10.10c have the same area as the original triangle, which proves the theorem. ∎

We now consider quadrilaterals. How does one find the area of a general quadrilateral? One method is presented in the following theorem.

Theorem 10.6. *The area of a convex quadrilateral Q is equal to one-half the area of a parallelogram P whose sides are parallel to and equal in length to the diagonals of Q.*

Proof. See Figure 10.11, where we have overlaid a grid formed by the diagonals of Q on the tiling of the plane by copies of Q. The result in the theorem is immediate. ∎

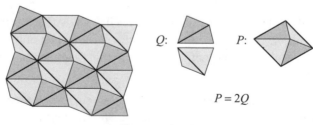

$Q:$ $P:$

$P = 2Q$

Figure 10.11.

The theorem also holds for non-convex quadrilaterals. See Challenge 10.5.

In Section 6.5 we encountered the Napoleon triangle—the equilateral triangle whose vertices were the centers of the three equilateral triangles erected on the sides of a general triangle. Is there a similar result for quadrilaterals and squares?

The answer is yes for parallelograms and squares, which is illustrated in the following theorem [Flores, 1997].

Theorem 10.7. *The quadrilateral determined by the centers of the squares externally erected on the four sides of an arbitrary parallelogram is a square.*

Proof. See Figure 10.12 for a portion of the plane tiling generated by a parallelogram and the squares erected on its sides and an overlay created by connecting the centers of the squares. ∎

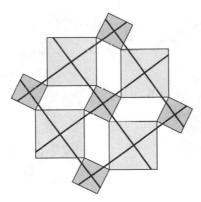

Figure 10.12.

If we let a and b denote the lengths of the sides of the parallelogram, P its area, and S the area of each overlay square, then we have the following parallelogram analog of Napoleon's theorem: $2S = 2P + a^2 + b^2$.

10.3 Infinitely many proofs of the Pythagorean theorem

In Chapter 5 we encountered two dissection proofs of the Pythagorean theorem, illustrated in Figure 5.2, which we reproduce here as Figure 10.13.

How are such dissections proofs created? One way to create them is to use a plane tiling based on squares. It is a simple manner to tile the plane with a collection of squares of two different sizes, as has been done for centuries

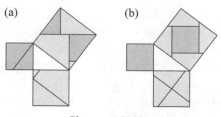

(a)   (b)

Figure 10.13.

in buildings. In Figure 10.14a we see the painting *Street Musicians at the Doorway of a House* by Jacob Ochtervelt (1634–1682). The tiling pattern on the floor in the house is also found in the Oberkapfenberg Castle in Styria, Austria (see Figure 10.14b), and is often called the *Pythagorean tiling*, for reasons we shall now see. It is not an edge-to-edge tiling.

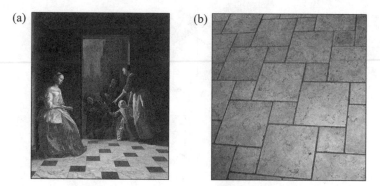

(a)   (b)

Figure 10.14.

In the Pythagorean tiling draw lines through the upper right hand corners of each of the smaller dark gray squares as shown in Figure 10.15a. This overlays on the tiling a grid of congruent transparent squares that we call the *hypotenuse grid*—note the right triangle formed in the lower left corner of the larger light gray squares. If we let a and b denote the legs and c hypotenuse of the right triangles, then the areas of the dark and light gray squares are a^2 and b^2, and the transparent squares formed by the hypotenuse grid illustrates the dissection proof that $c^2 = a^2 + b^2$ from Figure 10.13a

Now shift the hypotenuse grid so that the intersections of the lines lie at the centers of the larger light gray squares, as shown in Figure 10.15b. This yields the dissection proof of the Pythagorean theorem shown in Figure 10.13b.

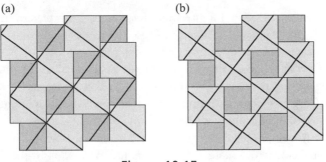

(a)   (b)

Figure 10.15.

Hence we can create as many different dissection proofs of the Pythagorean theorem as there are ways to overlay the hypotenuse grid on the Pythagorean tiling. Thus we have proved.

Theorem 10.8. *There are infinitely many different dissection proofs of the Pythagorean theorem.*

In fact, there are uncountably infinitely many such dissection proofs, and in each one the square of the hypotenuse is dissected into nine or fewer pieces!

Other plane tilings and overlay grids yield additional proofs of the Pythagorean theorem. See Challenge 10.1. The same idea (a plane tiling plus an overlay grid) underlies the Greek cross dissections in Section 8.5 (see Challenge 10.4). In Section 6.4 we encountered a plane tiling using an arbitrary triangle and the three equilateral triangles constructed on its sides and in Section 8.2 we gave a tiling proof that a square inscribed in a semicircle has $2/5$ the area of a square inscribed in a circle of the same radius.

Escher's techniques

The tiling patterns in the Alhambra palace in Granada, Spain are extremely rich, incorporating a variety of geometric shapes and vibrant colors, and their study has fascinated mathematicians among the numerous visitors.

While many visiting mathematicians were interested in finding symmetry groups among the Alhambra tilings, the Dutch artist Maurits Cornelius Escher (1898–1972) spent time during his visits in 1922 and 1936 creating drawings of the tiles [Schattschneider, 2004]. This was a fruitful visit for Escher, for afterward, inspired by these patterns he began applying the techniques of the Moorish artisans to generate more sophisticated tiles, using human and animal forms.

Escher would begin with a tiling based on a polygonal tile and modify the tile by an area-preserving transformation that preserves the tiling property. Among the transformations that accomplish this are removing a piece of the tile from one edge to add to another side and removing a piece from one-half of one side to add to the other half. These are illustrated below for square and triangular tiles, following by the tilings created by (a) translation, (b) reflection, and (c) rotation. These tilings are based on sketches drawn by Escher.

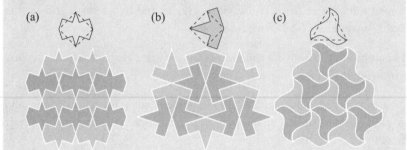

Figure 10.16.

Similar techniques we used by Escher on other polygonal tiles such as hexagons and parallelograms.

10.4　The leaping frog

In this section we will employ a tiling of the plane with *numbered* squares to solve a problem about a simple solitaire game. The game is "The Leaping Frog" ("La Rana Saltarina" in Spanish) described by Miguel de Guzmán [de Guzmán, 1997]. The board for the game is the plane divided into squares as on a chessboard, with a horizontal line L as shown in Figure 10.17. The player may then place as many coins as desired on the squares below the line. The rule of play is simple: coins are moved only by jumps to left, right, or upwards over a single adjacent coin (which is then removed from the board) to an adjacent vacant square. The player wins the game by bringing at least

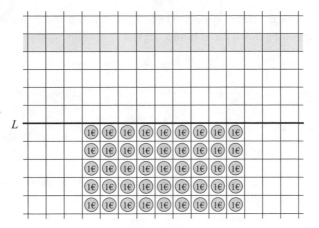

Figure 10.17.

one coin to the fifth row of squares (shaded gray) above the line L. The reader may wish to play the game before reading further.

We now show that it is impossible for the player to win. The argument we present [de Guzmán, 1997] is due to John Conway, and employs the golden ratio φ (see Section 2.4). Assume that it is possible to bring a coin to the fifth row of squares above L. Number each square in that row and all the rows below with powers of $\omega = \varphi - 1 = (-1 + \sqrt{5})/2$ as shown in Figure 10.18. Note that one square in the fifth row is assigned the value 1, $|\omega| < 1$, and that $\omega^2 = 1 - \omega$.

$\cdots$	$\cdots$	$\cdots$	ω^4	ω^3	ω^2	ω	1	ω	ω^2	ω^3	ω^4	$\cdots$	$\cdots$	$\cdots$
$\cdots$	$\cdots$	$\cdots$	ω^5	ω^4	ω^3	ω^2	ω	ω^2	ω^3	ω^4	ω^5	$\cdots$	$\cdots$	$\cdots$
$\cdots$	$\cdots$	$\cdots$	ω^6	ω^5	ω^4	ω^3	ω^2	ω^3	ω^4	ω^5	ω^6	$\cdots$	$\cdots$	$\cdots$
$\cdots$	$\cdots$	$\cdots$	ω^7	ω^6	ω^5	ω^4	ω^3	ω^4	ω^5	ω^6	ω^7	$\cdots$	$\cdots$	$\cdots$
$\cdots$	$\cdots$	$\cdots$	ω^8	ω^7	ω^6	ω^5	ω^4	ω^5	ω^6	ω^7	ω^8	$\cdots$	$\cdots$	$\cdots$
$\cdots$	$\cdots$	$\cdots$	ω^9	ω^8	ω^7	ω^6	ω^5	ω^6	ω^7	ω^8	ω^9	$\cdots$	$\cdots$	$\cdots$
$\cdots$	$\cdots$	$\cdots$	ω^{10}	ω^9	ω^8	ω^7	ω^6	ω^7	ω^8	ω^9	ω^{10}	$\cdots$	$\cdots$	$\cdots$
$\cdots$	$\cdots$	$\cdots$	ω^{11}	ω^{10}	ω^9	ω^8	ω^7	ω^8	ω^9	ω^{10}	ω^{11}	$\cdots$	$\cdots$	$\cdots$
$\cdots$	$\cdots$	$\cdots$	ω^{12}	ω^{11}	ω^{10}	ω^9	ω^8	ω^9	ω^{10}	ω^{11}	ω^{12}	$\cdots$	$\cdots$	$\cdots$
$\cdots$	$\cdots$	$\cdots$	ω^{13}	ω^{12}	ω^{11}	ω^{10}	ω^9	ω^{10}	ω^{11}	ω^{12}	ω^{13}	$\cdots$	$\cdots$	$\cdots$

Figure 10.18.

We now sum the (infinitely many) numbers assigned to the squares below the line, summing down the columns beginning with the center column:

$$S = (\omega^5 + \omega^6 + \cdots) + 2(\omega^6 + \omega^7 + \cdots) + 2(\omega^7 + \omega^8 + \cdots) + \cdots$$

$$= \frac{\omega^5}{1 - \omega} + 2\frac{\omega^6}{1 - \omega} + 2\frac{\omega^7}{1 - \omega} + \cdots = \omega^3 + 2\omega^4 + 2\omega^5 + \cdots$$

$$= (\omega^3 + \omega^4 + \omega^5 + \cdots) + (\omega^4 + \omega^5 + \cdots) = \frac{\omega^3}{1 - \omega} + \frac{\omega^4}{1 - \omega}$$

$$= \omega + \omega^2 = 1.$$

Therefore any finite sum of numbers below L will be strictly less than 1. When we begin with any finite number of coins below L, the sum of the numbers in the occupied squares will be less than 1. A jump in the direction of decreasing exponents will leave the sum unchanged, e.g., $\omega^7 + \omega^6 = \omega^5$ (see Figure 10.19a); whereas a jump in the direction of increasing exponents decreases the sum, e.g., $\omega^8 + \omega^7 = \omega^6 > \omega^9$ (see Figure 10.19b).

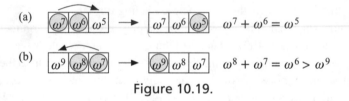

Figure 10.19.

Consequently the sum of the numbers assigned to the squares occupied by the coins begins with a value less than 1 and cannot increase, thus it is impossible to arrive at a value 1 in the fifth row above L.

10.5   The seven friezes

A *frieze* is a sculptured or ornamented horizontal band, such as can be found above columns in architecture, on wallpaper trim near the ceiling of a room, or on pieces of furniture. The design is a tiling wherein the tiles usually repeat and form a pattern, called a *frieze pattern*. Mathematicians classify the patterns by the symmetries they exhibit. Two patterns are said to be the same if they have the same *isometries*, i.e., distance-preserving transformations such as translations, rotations, and reflections.

A frieze always has translational symmetry, by definition. The reflections that leave the frieze invariant are reflections in a vertical line (a vertical reflection) or in a horizontal line (a horizontal reflection, in the horizontal line that bisects the frieze). The only nontrivial rotation that leaves the

Table 10.2.

1.	Ø	5.	v	9.	hr	13.	ghv
2.	g	6.	gh	10.	hv	14.	grv
3.	h	7.	gr	11.	rv	15.	hrv
4.	r	8.	gv	12.	ghr	16.	$ghrv$

frieze invariant is a 180° rotation. The final isometry is the glide reflection—translation horizontally followed by a horizontal reflection. Thus we have the following five types of symmetry: t—translation; v—vertical reflection; h—horizontal reflection; r—180° rotation; and g—glide reflection.

As every frieze has translation symmetry, there are $2^4 = 16$ cases to consider, depending on whether the frieze does or does not possess each of the other four types of symmetry. The 16 cases are listed in Table 10.2.

We will now show that nine of the sixteen cases are impossible, and each of the remaining seven exists as friezes, proving the following theorem. Most proofs of this theorem use group theory, but the following clever proof [belcastro and Hull, 2002] uses only combinatorial arguments.

Theorem 10.9. *There are exactly seven frieze patterns.*

Proof. In our proof we use the letter "p" as a "tile" to generate the frieze via the five isometries. We chose "p" since it does not exhibit any of the isometries under consideration.

We first show in Figure 10.20 that four of the cases (1, 2, 4, and 5) do indeed exist as friezes. In the figure (and those to follow) the arrow → indicates the pattern produced via translation. In each case note that no additional symmetries result in the frieze.

However, when we do the same with case 3 (a horizontal reflection h), we find that translation produces a frieze with glide reflection symmetry, as

(a) $\varnothing : \text{p} \quad \rightarrow \quad \text{p p p p p p}$

(b) $r : \text{p}_{\text{d}} \quad \rightarrow \quad \text{p} \ \text{p} \ \text{p} \atop \text{d} \ \text{d} \ \text{d}$

(c) $v : \text{p} | \text{q} \quad \rightarrow \quad \text{p q p q p q}$

(d) $g : \text{p}_{\text{b}} \quad \rightarrow \quad \text{p} \ \text{p} \ \text{p} \atop \text{b} \ \text{b} \ \text{b}$

Figure 10.20.

noted by the bold letters in Figure 10.21. So any time a frieze has horizontal symmetry, it must also have glide reflection symmetry. This eliminates cases 3 (h), 9 (hr), and 15 (hrv), and shows that case 6 (gh) does exist as a frieze.

$$h : \frac{\text{p}}{\text{b}} \quad \rightarrow \quad \begin{array}{cccccc} \text{p} & \text{p} & \textbf{p} & \text{p} & \text{p} & \text{p} \\ \text{b} & \text{b} & \textbf{b} & \textbf{b} & \text{b} & \text{b} \end{array}$$

Figure 10.21.

We now consider the symmetries h, g, r, and v in pairs. We first see, in Figures 10.22a and 10.22b, that if the frieze has h and v symmetry (in either order), then it also has r and g symmetry. Hence we can eliminate cases 10 (hv) and 13 (ghv), and confirm that case 16 ($ghrv$) exists. Similarly, as we see in Figures 10.22c and 10.22d, if a frieze has h and r symmetry, then it must also have v and g symmetry. Thus case 12 (ghr) must be eliminated.

(a) $h : \dfrac{\text{p}}{\text{b}}$ then $v : \dfrac{\text{p} \quad \text{q}}{\text{b} \quad \text{d}}$ $\rightarrow$ $\begin{array}{cccccc} \text{p} & \textbf{q} & \textbf{p} & \text{q} & \text{p} & \text{q} \\ \text{b} & \text{d} & \text{b} & \textbf{d} & \textbf{b} & \text{d} \end{array}$

(b) $v : \text{p} \,\big|\, \text{q}$ then $h : \dfrac{\text{p} \quad \text{q}}{\text{b} \quad \text{d}}$ $\rightarrow$ $\begin{array}{cccccc} \text{p} & \textbf{q} & \textbf{p} & \text{q} & \text{p} & \text{q} \\ \text{b} & \text{d} & \text{b} & \textbf{d} & \textbf{b} & \text{d} \end{array}$

(c) $h : \dfrac{\text{p}}{\text{b}}$ then $r : \dfrac{\text{p}}{\text{b}} \circ \dfrac{\text{q}}{\text{d}}$ $\rightarrow$ $\begin{array}{cccccc} \text{p} & \textbf{q} & \textbf{p} & \text{q} & \text{p} & \text{q} \\ \text{b} & \text{d} & \text{b} & \textbf{d} & \textbf{b} & \text{d} \end{array}$

(d) $r : \text{p} \circ \text{d}$ then $h : \dfrac{\text{p} \quad \text{q}}{\text{b} \quad \text{d}}$ $\rightarrow$ $\begin{array}{cccccc} \text{p} & \textbf{q} & \textbf{p} & \text{q} & \text{p} & \text{q} \\ \text{b} & \text{d} & \text{b} & \textbf{d} & \textbf{b} & \text{d} \end{array}$

Figure 10.22.

In considering the pair r, v of symmetries, we must be careful. If we first reflect, then we have two choices of location for the center of the rotation—on the line of reflection, or off it (and similarly if we rotate first). In Figures 10.23a and 10.23b, we see that if the center of rotation and the vertical line of symmetry coincide, then the result has both h and g symmetries. In Figures 10.23c and 10.23d, we see that if the center of rotation is off the vertical line of symmetry, then the result has g symmetry, but not h. consequently we can eliminate case 11 (rv), and confirm that case 14 (vrg) exists.

To summarize, at this point we have shown that seven cases in Table 10.2 exist as frieze patterns (1, 2, 4, 5, 6, 14, and 16) while seven cases (3, 9, 10, 11, 12, 13, and 15) are impossible. It remains only to examine cases 7 (gr) and 8 (gv). Note that since glide reflections are translational in nature, they can be performed only after all other transformations.

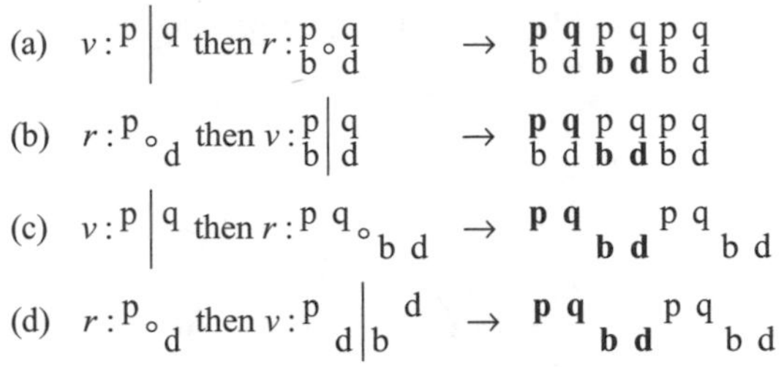

(a) $v: \begin{matrix} p \\ b \end{matrix} \Big| \begin{matrix} q \\ d \end{matrix}$ then $r: \begin{matrix} p \\ b \end{matrix} {}_{\circ} \begin{matrix} q \\ d \end{matrix}$ $\rightarrow$ $\begin{matrix} \mathbf{p} & \mathbf{q} & p & q & p & q \\ \mathbf{b} & \mathbf{d} & b & d & b & d \end{matrix}$

(b) $r: \begin{matrix} p \\ b \end{matrix} {}_{\circ} {}_{d}$ then $v: \begin{matrix} p \\ b \end{matrix} \Big| \begin{matrix} q \\ d \end{matrix}$ $\rightarrow$ $\begin{matrix} \mathbf{p} & \mathbf{q} & p & q & p & q \\ \mathbf{b} & \mathbf{d} & b & d & b & d \end{matrix}$

(c) $v: \begin{matrix} p \\ \end{matrix} \Big| \begin{matrix} q \\ \end{matrix}$ then $r: \begin{matrix} p & q \\ & b & d \end{matrix} {}_{\circ}$ $\rightarrow$ $\begin{matrix} p & q \\ & \mathbf{b} & \mathbf{d} \end{matrix} \begin{matrix} p & q \\ & b & d \end{matrix}$

(d) $r: \begin{matrix} p \\ b \end{matrix} {}_{\circ} {}_{d}$ then $v: \begin{matrix} p \\ & d \end{matrix} \Big| \begin{matrix} d \\ b \end{matrix}$ $\rightarrow$ $\begin{matrix} p & q \\ & \mathbf{b} & \mathbf{d} \end{matrix} \begin{matrix} p & q \\ & b & d \end{matrix}$

Figure 10.23.

(a) $r: \begin{matrix} p \\ b \end{matrix} {}_{\circ} {}_{d}$ then $g: \begin{matrix} p \\ & d & b \end{matrix} {}^{q}$ $\rightarrow$ $\begin{matrix} p \\ & d & b \end{matrix} {}^{q} \Big| \begin{matrix} p \\ & d & b \end{matrix} {}^{q}$

(b) $v: \begin{matrix} p \\ \end{matrix} \Big| \begin{matrix} q \\ \end{matrix}$ then $g: \begin{matrix} p & q \\ & b & d \end{matrix}$ $\rightarrow$ $\begin{matrix} p & q \\ & b & d \end{matrix} {}_{\circ} \begin{matrix} p & q \\ & b & d \end{matrix}$

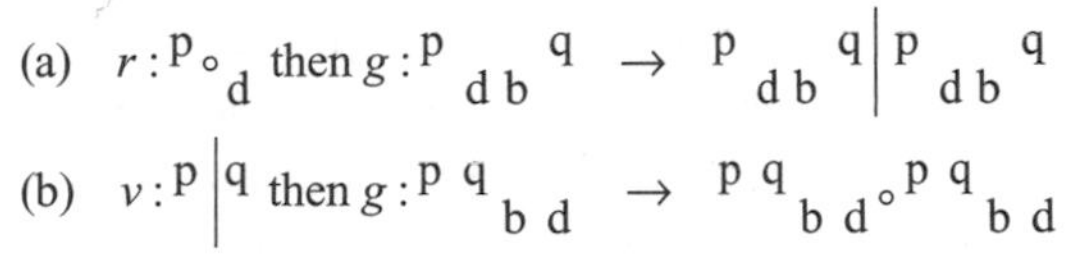

Figure 10.24.

In Figure 10.24a, we see that if a frieze has r and g symmetry, then it must also have v symmetry and in Figure 10.24b we see that if a frieze has v and g symmetry, it must also have r symmetry. Hence we eliminate cases 7 (gr) and 8 (gv) as frieze patterns, and the proof is complete. ∎

10.6 Colorful proofs

The use of color in a proof is a way to identify certain subsets of a given set in a simple visual manner. For example, coloring the vertices of triangles in the proof of the art gallery theorem in Section 4.6 enabled us to complete the proof quickly and elegantly. In addition to coloring points, we can color line segments or regions in the plane, such as tiles. We begin with a theorem whose proof uses colored line segments. [As you have no doubt noticed by now, this book does not contain any illustrations in color. Hence our "colors" must, unfortunately, be white, black, and shades of gray.]

Theorem 10.10. *Suppose in a gathering of six people, any pair of individuals are either acquainted ("friends") or unacquainted ("strangers"). Show that there must be a set of three (or more) individuals who are either mutual friends or mutual strangers.*

Proof. Represent the six individuals by the vertices of a regular hexagon, and join two vertices by a gray line (an edge or diagonal of the hexagon) if they are friends, and by a black line if they are strangers. See Figure 10.25a for an example.

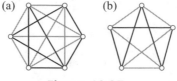

Figure 10.25.

We must show that there exists a triangle of lines all the same color (a *chromatic* triangle). Consider one vertex (say A) and the five lines joining it to the other five vertices. At least three must be of the same color (by the pigeonhole principle), say gray. Let $\{B, C, D\}$ be the other endpoints of those three lines. If one of the lines connecting two vertices in $\{B, C, D\}$ is gray, we have a gray triangle. If not then all three lines joining pairs from $\{B, C, D\}$ are black, and we have a black triangle. Figure 10.25b shows that "six" cannot be replaced by "five" in Theorem 10.10. ■

Our next examples concern tiles called *polyominoes*, a generalization of dominoes. A polyomino is a union of unit squares in which each square shares at least one edge with another square. See Figure 10.26 for illustrations of a domino, two types of *trominoes* (straight and L), and five types of *tetrominoes* (straight, square, T, skew, and L). Polyominoes, like dominoes, can be rotated and turned over, and are still considered to be of the same type. Our examples concern tiling rectangular chessboards whose components are unit squares.

We first consider tiling m-by-n chessboards with dominoes. The standard coloring of the chessboard is all one needs to see that the board can be tiled if and only if mn is even. So we now consider *deficient* chessboards (with

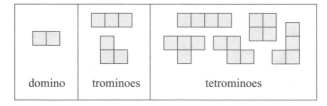

Figure 10.26.

the standard coloring), chessboards where one or more squares have been removed. If, on an *m*-by-*n* board with *mn* even, we remove two squares of the same color (e.g., opposite corners of an 8-by-8 board), then the board cannot be tiled, as we no longer have the same number of squares of each color. But what happens if we remove a pair of squares with different colors from an *m*-by-*n* board with *mn* even? What if we remove one square of the majority color (i.e., the color of one of the corner squares) from an *m*-by-*n* board with *mn* odd? Can we then tile the deficient chessboards?

Theorem 10.11. *A deficient m-by-n chessboard with the standard coloring, m and n each at least 2, can be tiled with dominoes if* (i) *mn is even and any two squares of different colors are removed, or* (ii) *mn is odd and any one square of the majority color is removed.*

Proof. Part (i) of this theorem is known as *Gomory's theorem*. Its proof first appeared in [Gardner, 1962]. In the proof we partition the chessboard into a closed path one square wide, as indicated by the heavy lines in Figure 10.27a, where the horizontal side of the board has an even number of squares.

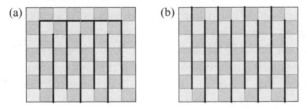

Figure 10.27.

Removing any two squares of different colors will cut the path into two segments (or one is the two squares are adjacent), each segment will contain an even number of squares of alternating colors, and hence can be tiled by dominoes.

When *mn* is odd, we create an open-ended path [Trigg, 1973] as illustrated in Figure 10.27b. When one square of the majority color (here light gray) is removed, the result is two segments each with an even number of squares (or one segment if a square in the upper left or lower right corner is removed) of alternating colors, which can be tiled with dominoes.

The case where *m* or *n* equals 1 requires an additional condition. See Challenge 10.7. ■

We now consider tiling standard and deficient 8-by-8 chessboards with different types of trominoes and tetrominoes. Obviously the 8-by-8 board cannot be tiled by trominoes (64 is not a multiple of 3), so we remove one square. Can the resulting deficient board be tiled with straight trominoes? With L-trominoes?

Theorem 10.12. *An* 8-*by*-8 *chessboard with one square removed can be tiled by straight trominoes if and only if the square removed is one of the four black squares in Figure* 10.28a.

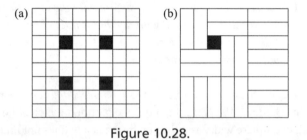

Figure 10.28.

Proof. [Golomb, 1954]. We color the board with three colors so that a straight tromino placed on the board, horizontally or vertically, will cover squares of three different colors. One way to do this is illustrated in Figure 10.29a. Since there are 21 white squares, 21 light gray squares, and 22 dark gray squares, the square removed must be one of the dark gray squares. But the same is true for the coloring in Figure 10.29b, hence it is necessary to remove a square colored dark gray in both figures, which are precisely the four black squares in Figure 10.28a. Figure 10.28b (and its rotations) shows that this condition is also sufficient. ∎

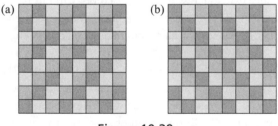

Figure 10.29.

We now turn to tiling the 8-by-8 board with tetrominoes. Figure 10.30a [Golomb, 1954] shows that square, straight, L, and T tetrominoes tile the board. However, skew tetrominoes do not (see Challenge 10.8). What about

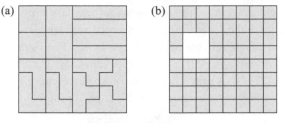

Figure 10.30.

deficient boards? Clearly we need to remove four squares, so we consider a board with the standard coloring with a 2-by-2 hole (with no restriction on where the hole is located), as in Figure 10.30b.

Theorem 10.13. *An* 8-*by*-8 *chessboard with a* 2-*by*-2 *hole* (i) *cannot be tiled by T-tetrominoes,* (ii) *cannot be tiled by L-tetrominoes, and* (iii) *cannot be tiled by any combination of straight and skew tetrominoes.*

Proof [Golomb, 1954]. In this proof we use three different colorings of the board. For (i) we use the standard coloring (as in Figure 10.27). The square hole removes two squares of each color, hence we have 30 light gray and 30 dark gray squares in the deficient board. Each T-tetromino will cover either one light gray and three dark gray squares or one dark gray and three light gray squares. Suppose x tetrominoes cover one light gray and three dark gray squares and y tetrominoes cover one dark gray and three light gray squares. Then $x + 3y = y + 3x$ (i.e., $x = y$) and $x + y = 15$, thus $2x = 15$, which is impossible.

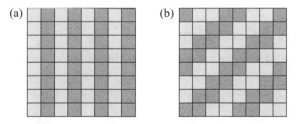

Figure 10.31.

For (ii), we use the column-wise coloring in Figure 10.31a. Each L-tetromino will cover either one light gray and three dark gray squares or one dark gray and three light gray squares, and so the same argument applies here so that this tiling is impossible as well.

For (iii) we use the diagonal coloring in Figure 10.31b. There are 32 light gray and 32 dark gray squares. But here the 2-by-2 square hole will remove either one light gray and three dark gray, or one dark gray and three light gray squares; in either case the deficient board with have an odd number of light gray and an odd number of dark gray squares. Each straight tetromino, no matter its orientation, will cover an even number (2) of light gray and an even number (2) of dark gray squares; and each skew tetromino, no matter its orientation, will cover an even number (0, 2, or 4) of light gray and an even number (4, 2, or 0) of dark gray squares. Hence any collection of straight and skew tetrominoes will cover an even number of light gray and an even number of dark gray squares, so a tiling is impossible. ■

For more results about tiling with polyominoes and some colorful proofs, see [Golomb, 1965].

Up to now our chessboards have been either squares or rectangles. Suppose we construct an 8-by-8 chessboard with the standard coloring from an arbitrary convex quadrilateral, as shown in Figure 10.32. Rather than try to tile this board, we ask: is the sum of the areas of the 32 light gray cells the same as the sum of the areas of the 32 dark gray cells?

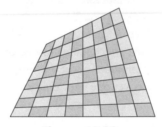

Figure 10.32.

We begin with a simpler 2-by-2 chessboard.

Theorem 10.14. *Create a 2-by-2 chessboard from a convex quadrilateral by joining midpoints of opposites sides and coloring the resulting cells as shown in Figure* 10.33a. *Then the areas of the light gray and dark gray cells are the same.*

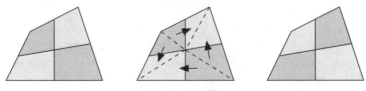

Figure 10.33.

Proof. Draw line segments joining the vertices of the quadrilateral to the point of intersection of the lines connecting the midpoints of the sides, as shown in Figure 10.33b. This creates four triangles with medians that partition each triangle into two parts with the same area, and the result follows. ∎

The result in Theorem 10.14 immediately extends to any $2n$-by-$2n$ convex quadrilateral chessboard formed joining equidistant points on the sides, as shown in Figure 10.34, to yield the following corollary.

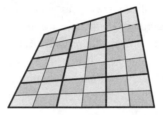

Figure 10.34.

Corollary 10.2. *Create a $2n$-by-$2n$ chessboard from a convex quadrilateral by joining an odd number of equidistant points on opposites sides and coloring the resulting cells as shown in Figure 10.34. Then the areas of the light gray and dark gray cells are the same.*

Another tiling problem with a colorful solution concerns *calissons*— French sweets in the shape of two equilateral triangles joined on an edge (see Figure 10.35a). Calissons could come in a hexagonal box (but apparently do not), packed as illustrated in Figure 10.35b with the short diagonal of each calisson parallel to one of the three sides of the box, so that there are three possible orientations for each calisson.

(a)

(b)

Figure 10.35.

If we place a triangular grid on the hexagonal box, then each calisson is a domino for this grid, covering two equilateral triangles. The next theorem tells us how many calissons in the box have each of the three orientations when the box is full of (i.e., tiled by) calissons.

Theorem 10.15. *In any packing of calissons in a regular hexagonal box, the number of calissons in each of the three orientations is the same and equal to one-third of the number of calissons in the box.*

Proof [David and Tomei, 1989]. In Figure 10.36a we see an arbitrary filling of the hexagonal box with calissons, and in Figure 10.36b we have colored the calissons in the three different orientations with different colors. Once the calissons have been colored, they appear as a collection of cubes in the corner of a room with a square floor and square walls. Looking at this configuration from above one sees just the tops of the cubes, which of course cover the floor; the same is true if one views the configuration from one of the sides. Hence the number of cube faces, i.e., calissons, in each orientation is the same, which proves the theorem.   ■

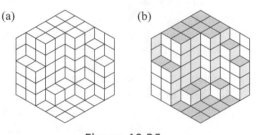

Figure 10.36.

We conclude this section with a topic that may have inspired much of the mathematical theory of coloring—maps. A mathematical *map* is a partition of the plane into a finite number of regions (states or countries) whose boundaries are a finite number of line segments or simple curves (i.e., there are no self-intersections). The object is to color the map using as few colors as possible in such a way that adjacent countries have different colors. Two countries are adjacent if they share a common boundary other than a single point.

The *Four Color Theorem* states that four colors suffice to color any map. While the Four Color Theorem is charming, its proof is not, so we shall consider only three simpler theorems.

The Four Color Theorem

In 1852 Francis Guthrie, a student in Edinburgh, observed that it seemed to be true that four colors suffice to color a map with the condition that no two adjacent countries receive the same color. Francis then asked his brother Frederick "Are four colors always sufficient?" and Frederick asked his mathematics teacher in London, Augustus De Morgan, and interest in the problem spread throughout Europe. The Four Color Conjecture became widely known after 1878 when Arthur Cayley admitted he had been unable to find a proof.

In 1879 Sir Alfred Bray Kempe published a proof, but eleven years later Percy John Heawood discovered a fatal flaw in Kempe's proof. Subsequent attempts at a proof led to many developments in graph theory, but no proof of the conjecture. Finally, in 1976 Wolfgang Haken and Kenneth Appel of the University of Illinois proved the conjecture, with a proof that required many hours of computer time to check a multitude of cases. In recognition of this achievement, the postmark of the Department of Mathematics at Illinois was changed to read "Four colors suffice," as seen in Figure 10.37.

Figure 10.37.

In spite of the fascination of this problem in the mathematical community for over a century, it seems to have been of little interest to cartographers.

The Three Color Theorem 10.16. *Some maps require more than three colors.*

Proof. See Figure 10.38 for a proof with a mathematical map; see Challenge 10.13 for proofs with real-world maps. ∎

The Two Color Theorem 10.17. *Any map determined by a finite collection of straight lines and circles can be colored with two colors.*

Proof. While this theorem can be proven easily by induction, we present a direct proof, adapted from the one in [Gardner, 1971]. Assign each line and

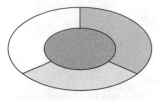

Figure 10.38.

circle a direction or orientation, as indicated by the arrows in Figure 10.39a. For each region R in the map, let $f(R)$ be the number of lines and circles such that R is on the right hand side of the line or circle, as shown in Figure 10.39b. If $f(R)$ is even, color R white; if $f(R)$ is odd, color R gray, as in Figure 10.39c. This is a two coloring such that regions sharing a common boundary have different colors. ■

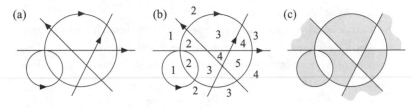

Figure 10.39.

The Three Dimensional Four Color Theorem 10.18. *In space, four colors are not enough.*

Proof. See Figure 10.40, where we show that no finite number of colors is sufficient to color a three-dimensional map [Alsina and Nelsen, 2006]. ■

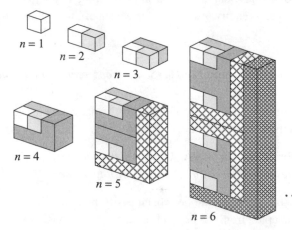

Figure 10.40.

10.7 Dodecahedra and Hamiltonian circuits

The dodecahedron is one of the Platonic solids, composed of twelve congruent regular pentagons (see Figure 10.41a). It has 20 vertices and 30 edges. In 1857 Sir William Rowan Hamilton (1805–1865) asked if it were possible to find a path along the edges of a dodecahedron that passes through each vertex once and only once and returns to the initial vertex. Such a path (if it exists) is called a *Hamiltonian circuit.*

In Figure 10.41b we see such a circuit on the graph of the dodecahedron, where we have projected the faces, edges, and vertices onto the plane. In fact, one can show that each of the five Platonic solids (tetrahedron, cube, octahedron, dodecahedron, and icosahedron) has a Hamiltonian circuit.

(a)   (b)

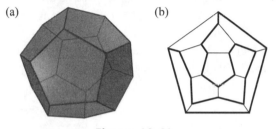

Figure 10.41.

While this problem was easy to solve, in general, proving that Hamiltonian circuits exist or do not exist for a given polyhedron or graph can be challenging. However, occasionally such a problem has a charming solution. Here is one.

The *rhombic dodecahedron* is a solid with twelve congruent rhombi as its faces. See Figure 10.42. It has 14 vertices and 24 edges. Does it have a Hamiltonian circuit?

Theorem 10.19. *There is no Hamiltonian circuit on the rhombic dodecahedron.*

(a)   (b)

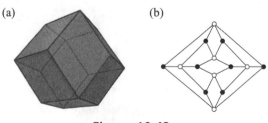

Figure 10.42.

Proof (attributed to H. S. M. Coxeter in [Gardner, 1971]).   Some of the vertices of the rhombic dodecahedron are of degree three (they are the endpoints of three edges), while others have degree four. Color the degree three vertices black and the degree four vertices white, as in the planar projection of the rhombic dodecahedron in Figure 10.41b. Each edge joins a black vertex to a white one. So if a Hamiltonian circuit exists, the vertices will alternate in color along the path. However, there are eight black and six white vertices, so no Hamiltonian circuit exists. ■

A colorful solution to a Lewis Carroll challenge

Charles Lutwidge Dodgson, better known as Lewis Carroll, author of the *Alice* books (see Section 5.11) was fond of asking children mathematical puzzles, such as the following [Gardner, 1971]: Can you draw the design in Figure 10.42a on a sheet of paper without lifting the pencil from the paper? This is easily done if intersections are permitted, but more difficult if intersections are not allowed.

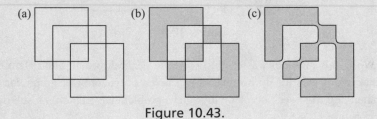

Figure 10.43.

T. H. O'Beirne of Glasgow devised the colorful solution illustrated in Figures 10.42b and 10.42c: color the regions with two colors, and then break them apart at certain vertices in such a way to leave a single simple region in one of the colors. The desired path is then the boundary of this region.

10.8   Challenges

10.1 Floor tiles in the Salon de Carlos V in the Real Alcázar of Seville, Spain (Figure 10.44a) suggest another tiling proof of the Pythagorean theorem. Use the tiling with rectangles and squares and the overlay grid of squares (Figure 10.44b) to prove the theorem.

(a)   (b)

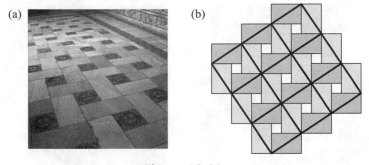

Figure 10.44.

10.2 In the regular hexagonal tiling, identify one hexagonal tile as h, and consider the hexagon H determined by the centers of the six hexagons surrounding h.

    (a) Show that the area of H is three times the area of h.

    (b) Show that h and the six surrounding hexagons can be partitioned into six pentagons to create a pentagonal tiling. This tiling is sometimes known as the *floret pentagonal tiling*, as the sets of six pentagons resemble the petals of a flower.

10.3 Show that a convex pentagon with two sides parallel tiles the plane.

10.4 (a) Show that a Greek cross composed of five congruent unit squares (see Section 8.5) tiles the plane.

    (b) Overlay the tiling in (a) with square grids to obtain the two dissections in Figure 8.20 (Hint: the side lengths of the squares in the overlay grid should be $\sqrt{5}$.)

    (c) Overlay the tiling in (a) with square grids to obtain the two dissections in Figure 8.21 (Hint: the side lengths of the squares in the overlay grid should be $\sqrt{10}$.)

    (d) Show that the crosses in the Greek cross tiling can be partitioned to form a monohedral pentagonal tiling.

10.5 Prove Theorem 10.6 for non-convex quadrilaterals.

10.6 The friezes in Figure 10.45 are from various locations in Seville, Spain (the Real Alcázar, the Plaza de España, and the Parque de María Luisa). What symmetries are present in each? Disregard the colors of the tiles.

10.7 State and prove a theorem about tiling deficient 1-by-$2k$ chessboards with dominoes.

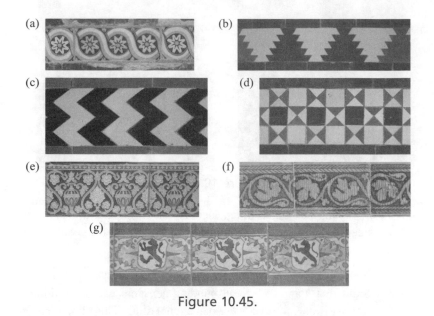

Figure 10.45.

10.8 Prove that a rectangular chessboard can never be tiled by skew tetro-minoes.

10.9 Prove that if an m-by-n rectangle can be tiled with L-tetrominoes, then 8 divides mn,

10.10 Solve Problem 1 from the 1976 U.S.A. Mathematical Olympiad:

(a) Suppose that each square on a 4×7 chessboard (Figure 10.46) is colored black or white. Prove that with any such coloring, the board must contain a rectangle (formed by the horizontal and vertical lines of the board) such as the one outlined in the Figure, whose four distinct corner squares are all of the same color.

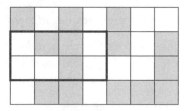

Figure 10.46.

(b) Exhibit a black-white coloring of a 4 × 6 chessboard in which the four corner squares of every such rectangle described above are not the same color.

10.11 Consider a circle with circumscribed and inscribed hexagons and the inscribed star hexagon, as shown in Figure 10.47. Show that the areas of the three polygons are in the ratio 4:3:2 [Trigg, 1962].

Figure 10.47.

10.12 How many colors does it take to color a crease pattern for flat origami?

10.13 Show that (a) a map of the 48 contiguous United States requires four colors, and (b) the same is true of a map of the European Union.

10.14 Is it possible to trace the design in Figure 10.48 with a single non-intersecting path?

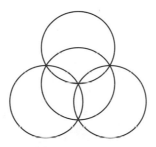

Figure 10.48.

CHAPTER 11

Geometry in Three Dimensions

Considerable obstacles generally present themselves to the beginner, in studying the elements of Solid Geometry, from the practice ... of never submitting to the eye of the student, the figures on whose properties he is reasoning.

Augustus De Morgan

The ludicrous state of solid geometry made me pass over this branch.

Plato, *The Republic* (VII, 528)

Space is almost infinite. As a matter of fact, we think it is infinite.

Dan Quayle
*44th Vice President of
the United States of America*

The geometry of three-dimensional Euclidean space is sometimes called solid geometry, since it has traditionally dealt with solids such as spheres, cylinders, cones, and polyhedra, as well as lines and planes in space. Three-dimensional space is the space we live in, and it is the setting for some lovely theorems and charming proofs.

In this chapter we present three different types of theorems and their proofs. We first consider three-dimensional versions of some two-dimensional theorems (such as the Pythagorean theorem). Second, we look at two-dimensional theorems whose proofs are surprisingly simple when the theorem and its proof are viewed from a three-dimensional perspective. Then we consider some classic theorems about polyhedra, the gems of three-dimensional space.

From Flatland to the Celestial Region

When Edwin Abbott Abbott (1838–1926) published his celebrated *Flatland: A Romance of Many Dimensions* [Abbott, 1884], he included the following dedication that remains today a suggestive invitation to three (and higher) dimensional geometry:

To

The Inhabitants of SPACE IN GENERAL

And H. C. IN PARTICULAR

This Work is Dedicated

By a Humble Native of Flatland

In the Hope that

Even as he was Initiated into the Mysteries

Of THREE Dimensions

Having been previously conversant

With ONLY Two

So the Citizens of that Celestial Region

May aspire yet higher and higher

To the Secrets of FOUR FIVE OR EVEN SIX Dimensions

Thereby contributing

To the Enlargement of THE IMAGINATION

And the possible Development

Of that most rare and excellent Gift of MODESTY

Among the Superior Races

Of SOLID HUMANITY

11.1   The Pythagorean theorem in three dimensions

Some consider the three dimensional version of the Pythagorean theorem to be the expression for the length of the diagonal of a rectangular box in terms of the lengths of the edges. Others maintain that the real extension should be some relationship between the areas of the faces of a *right tetrahedron* (a tetrahedron with three faces perpendicular to one another at one vertex), a three-dimensional analog of a right triangle.

We now state and prove one of these results. If one considers the three mutually perpendicular faces of a right tetrahedron to be its legs and the other face its hypotenuse, the theorem states that the square of the hypotenuse of a right tetrahedron equals the sum of the squares of the legs. The theorem is sometimes called *de Gua's theorem*, for the French mathematician Jean Paul de Gua de Malves (1713–1785).

de Gua's Theorem 11.1. *In a right tetrahedron, the square of the area of face opposite the vertex common to the three mutually perpendicular faces is equal to the sum of the squares of the areas of the other three faces.*

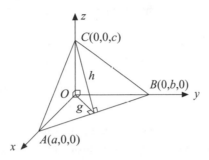

Figure 11.1.

Proof. Let $O = (0,0,0)$, $A = (a,0,0)$, $B = (0,b,0)$, and $C = (0,0,c)$ (with a,b,c positive) be the vertices of the right tetrahedron (See Figure 11.1). Let h denote the altitude to side AB in $\triangle ABC$, and let g denote its orthogonal projection onto the x-y plane. As such, g is the altitude to side AB in $\triangle AOB$. Then $g = ab/\sqrt{a^2 + b^2}$ (since the area of $\triangle AOB$ is both $ab/2$ and $g\sqrt{a^2 + b^2}/2$) and thus $h^2 = g^2 + c^2 = a^2b^2/(a^2 + b^2) + c^2$. If K denotes the area of $\triangle ABC$, then

$$K^2 = \left(\frac{|AB|h}{2}\right)^2 = \frac{1}{4}(a^2 + b^2)\left(\frac{a^2b^2}{a^2 + b^2} + c^2\right)$$

$$= \frac{1}{4}(a^2b^2 + a^2c^2 + b^2c^2) = \left(\frac{1}{2}ab\right)^2 + \left(\frac{1}{2}ac\right)^2 + \left(\frac{1}{2}bc\right)^2. \quad \blacksquare$$

11.2 Partitioning space with planes

In Heinrich Dörrie's classic book *100 Great Problems of Elementary Mathematics* [Dörrie, 1965], one of the eight problems attributed to Jakob Steiner is Problem 67:

> What is the maximum number of parts into which a space can be divided by n planes?

Our solution to this problem follows an argument presented by George Pólya over forty years ago. George Pólya (1887–1985) was renowned not only for his mathematics but also for his contributions to the heuristics of problem solving. In 1966 he filmed a session with future mathematics teachers, entitled *Let Us Teach Guessing*, leading them to discover the answer to Steiner's question. The film is now a DVD [Pólya, 1966] distributed by the Mathematical Association of America. Following Pólya's approach, we begin with n points on a line, then n lines in a plane, and finally n planes in space.

Clearly n distinct points partition the real line into $n+1$ intervals, and this simple observation serves to establish our first theorem.

Theorem 11.2. *The maximum number $P(n)$ of regions determined by n lines in the plane is given by $P(n) = 1 + n(n + 1)/2$.*

Proof. First note that the maximum number of regions will occur when no two lines are parallel and there is no point common to three or more lines. Clearly $P(0) = 1$, $P(1) = 2$ and $P(2) = 4$. Suppose $k - 1$ lines partition the plane into $P(k - 1)$ regions, and we add a new line to create as many additional regions as possible. Such a line will intersect all $k - 1$ of the other lines in $k - 1$ distinct points, partitioning the line into k intervals and each interval corresponds to a new region in the plane. See Figure 11.2, where we illustrate the $k = 4$ case, a line intersecting three lines to create four new regions in the plane.

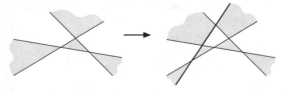

Figure 11.2.

Hence $P(k) = P(k - 1) + k$. Transposing the $P(k - 1)$ term on the right side to the left and summing yields

$$P(n) - P(0) = \sum_{k=1}^{n} [P(k) - P(k - 1)] = \sum_{k=1}^{n} k = \frac{n(n + 1)}{2},$$

and consequently $P(n) = 1 + n(n + 1)/2$ as claimed.   ■

In the next proof it will be advantageous to recall the triangular numbers $t_n = n(n + 1)/2$ (see Section 1.1) and to write $P(n)$ as $1 + t_n$.

Theorem 11.3. *The maximum number $S(n)$ of regions determined by n planes in space is given by $S(n) = (n^3 + 5n + 6)/6$.*

Proof. The maximum number of regions will occur when no two planes are parallel, there are no two parallel intersection lines, and there is no point common to four or more planes. Clearly $S(0) = 1$, $S(1) = 2$, $S(2) = 4$, and $S(3) = 8$. Suppose $k - 1$ planes partition space into $S(k - 1)$ regions, and we add a new plane to create as many additional regions as possible. Such a plane will intersect all $k - 1$ of the other planes, and those lines of intersection points partition the new plane into $P(k - 1)$ plane regions and each of those plane regions corresponds to a new region in space. Hence $S(k) = S(k - 1) + P(k - 1)$. Transposing the $S(k - 1)$ term on the right side to the left and summing yields

$$S(n) - S(0) = \sum_{k=1}^{n} [S(k) - S(k - 1)] = \sum_{k=1}^{n} P(k - 1)$$
$$= \sum_{k=1}^{n} (1 + t_{k-1}) = n + \frac{(n - 1)n(n + 1)}{6} = \frac{n^3 + 5n}{6},$$

where we have used the result in Theorem 1.8 to evaluate the final sum. Thus $S(n) = (n^3 + 5n + 6)/6$ as claimed. ∎

11.3 Corresponding triangles on three lines

Suppose two triangles ABC and $A'B'C'$ have their corresponding vertices on three lines that intersect at a point, as shown in Figure 11.3 Then we have

Theorem 11.4. *The three points of intersection (if they exist) of the pairs of lines determined by corresponding pairs of sides of triangles ABC and $A'B'C'$ in Figure 11.3 lie on a straight line.*

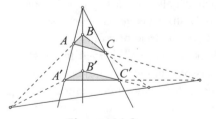

Figure 11.3.

Proof. Look at the three lines as being the edges of a triangular pyramid, or as the legs of a tripod. Then each of the two triangles is the intersection of the pyramid or tripod with a plane. The lines determined by corresponding pairs of sides of the triangles lie in those two planes, and unless the two planes are parallel, they will intersect in a straight line. ∎

11.4   An angle-trisecting cone

In 1896 Aubry created the following three-dimensional solution to the angle-trisection problem [Eves, 1983]. On a piece of paper draw a circle C of radius r with a central angle θ to be trisected, as illustrated in Figure 11.4a. Take another sheet of heavy paper or light cardboard, and cut out a 120° sector of a circle of radius $3r$, as illustrated in Figure 11.4b.

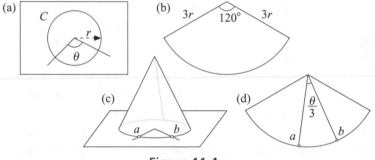

Figure 11.4.

Now transform the circular sector into a cone, and place it on the first sheet of paper so that the circle C is the base of the cone (see Figure 11.4c), and mark on the lateral surface of the cone the points a and b determined by the angle θ. Now open the cone and lay it flat, connect a and b to the vertex of the sector, and the resulting angle measures $\theta/3$ (see Figure 11.4d).

This conical approach also enables us to solve another classical problem unsolvable using only straightedge and compass: inscribing a regular heptagon in a circle. Given a circle of radius r, draw on heavy paper or light cardboard a circle of radius $8r/7$ divided into eight equal parts (see Figure 11.5). Using scissors cut out the circle, discard one slice, and form the remainder into a cone. Since the circumference of the cone is the same as the circumference of the circle, the cone fits exactly on top of the circle and divides it into seven equal parts, thus inscribing a regular heptagon in the circle.

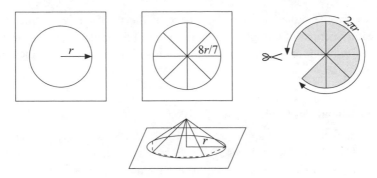

Figure 11.5.

11.5 The intersection of three spheres

If three spheres intersect one another in space, what is the set of points common to all three?

Theorem 11.5. *Three spheres (with non-collinear centers) intersecting one another have at most two points in common.*

Proof. First consider just two spheres. They intersect in a circle (or a point if they are tangent). This circle (or point) intersects the third sphere in two (or one) points. ∎

A beautiful consequence is the following result about circles in the plane [Bogomolny, 2009].

Theorem 11.6. *The three chords determined by the intersection of three circles (with non-collinear centers) meet in a point.* See Figure 11.6.

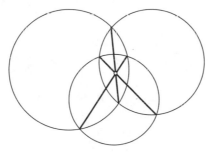

Figure 11.6.

Proof. Consider Figure 11.6 embedded in space, where the three circles are the equators of three spheres cut by the plane of this page. Then the three chords are the projections onto the page of the three circles of the pair-wise intersections of the spheres. By Theorem 11.5 the three spheres meet in two points, both of which project onto the point of intersection of the three chords. ■

The Global Positioning System (GPS)

The geometrical principle underlying the GPS is precisely Theo-rem 11.5 about three spheres. To locate an object equipped with a GPS unit three satellites are required. The GPS unit receives signals from each satellite, and thus knows the location of and distance to each satel-lite. Hence the GPS unit is located on three spheres, each one centered at one of the satellites, and consequently at one of the two points of inter-section of the spheres. If we know the approximate location of the GPS unit (e.g., on the earth's surface), we can discard one point. Appropriate software using elementary linear algebra enables the computation of the location of the GPS unit on the earth's surface.

The best viewing point

Where should one stand in front of a picture drawn in three-point per-spective to obtain the "best" view of the picture? If a cube is well drawn, to see it as a cube we want the angles of the faces to appear as right an-gles. Hence our eye needs to be on each of the three spheres that have as their diameters the line segments joining pairs of vanishing points. These three spheres intersect in two points, but only one is in front of the picture. See Figure 11.7.

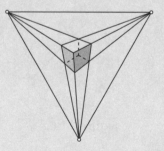

Figure 11.7.

11.6 The fourth circle

In 1916 R. A. Johnson discovered the following result [Johnson, 1916], which has been described as one of the few recent "really pretty theorems at the most elementary level of geometry" [Honsberger, 1976]. The elegant proof simply requires us to look at the configuration from a three dimensional point of view.

Theorem 11.7. *If three circles with the same radius r are drawn through a point P, then the other three points of intersection A, B, and C determine a fourth circle with the same radius.* See Figure 11.8a.

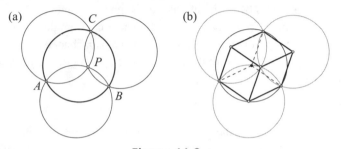

Figure 11.8.

Proof. The three points A, B, C of intersection and the centers of the circles form a hexagon divided into three rhombi, as illustrated in Figure 11.8b. The nine dark line segments each have length r, and drawing the three dashed line segments of length r produces a plane projection of a cube. Hence A, B, and C lie at a distance r from another point, and hence the fourth circle also has radius r. ■

11.7 The area of a spherical triangle

In plane geometry, knowing the size of the angles of a triangle tells us nothing about the area of the triangle but in spherical geometry it (along with the radius of the sphere) tells us everything.

In spherical geometry, "lines" are arcs of *great circles*, which are the intersections of the sphere with planes through the center of the sphere. In Figure 11.9a we see two great circles that intersect in antipodal (opposite) points to form four *spherical lunes*, one of which is shaded. Angles in lunes are *dihedral angles*, the angles between the two planes of the great circles. If the dihedral angle of a lune is θ (in this section we measure angles in radians)

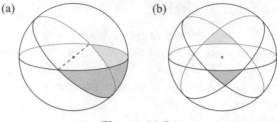

Figure 11.9.

and the radius of the sphere is r, then the area $L(\theta)$ of the lune is the fraction $\theta/2\pi$ of the surface area $4\pi r^2$ of the sphere, or $L(\theta) = 2r^2\theta$. Each pair of great circles produces two pairs of congruent spherical lunes.

Three great circles intersect to form eight spherical triangles, which come in congruent pairs. In Figure 11.9b we see such a pair, shaded dark and light gray. If the (dihedral) angles of the triangle are α, β, and γ, and its area is T, then we have

Theorem 11.8. $T = r^2(\alpha + \beta + \gamma - \pi)$.

Proof. Since total area of the six lunes determined by the three great circles is the surface area of the sphere plus four times the area of the spherical triangle, we have $4\pi r^2 + 4T = 2[L(\alpha) + L(\beta) + L(\gamma)] = 4r^2(\alpha + \beta + \gamma)$, and the result follows. ∎

The result in this theorem can be easily extended to spherical polygons. We triangulate the spherical n-gon into $n - 2$ spherical triangles, in a manner analogous to triangulation of plane n-gons. Applying the theorem to the resulting spherical triangles yields, for the area A_n of a spherical polygon, $A_n = r^2[S_n - (n - 2)\pi]$, where S_n denotes the sum (in radians) of the dihedral angles of the spherical n-gon. We will use this result in the next section.

11.8   Euler's polyhedral formula

> *The first important notions in topology were acquired in the course of the study of polyhedra.*
>
> Henri Lebesgue

In 1752 Leonhard Euler (1707–1783) announced his discovery of the remarkable formula $V - E + F = 2$, where V, E, and F are, respectively, the number of vertices, edges, and faces in a convex polyhedron. Euler and this formula were honored in 2007 when Switzerland issued a postage stamp

Figure 11.10.

to commemorate the 300th anniversary of Euler's birth (Figure 11.10a). The polyhedron illustrated on the stamp is sometimes called *Dürer's solid*, since it appears in the engraving *Melencolia I* (Figure 11.10b) created by Albrecht Dürer in 1514. For Dürer's solid, $V = 12$, $E = 18$, $F = 8$, and $12 - 18 + 8 = 2$.

The formula is remarkable since in applies to any convex polyhedron, regardless of its size or shape. The first rigorous proof of the formula appears to be one published in 1794 by Adrien-Marie Legendre (1752–1833), which we present next. Legendre's proof is based on the spherical geometry discussed in the previous section. Many other proofs are known; see [Eppstein, 2005] for a collection of nineteen different proofs.

Theorem 11.9 (Euler's polyhedron formula). *If V, E, and F are, respectively, the number of vertices, edges, and faces in a convex polyhedron, then* $V - E + F = 2$.

Proof. We begin by scaling and inserting the *skeleton* of the polyhedron (the network of vertices and edges without the faces, as illustrated on the stamp) into a sphere of radius 1, and project the skeleton onto the sphere by means of a light source at the sphere's center. This radial projection creates a spherical polygon whose edges are arcs of great circles and with exactly the same values of V, E, and F as the original polyhedron. Since the sphere has radius 1 its surface area is 4π, which is also the sum of the areas of the F spherical polygon faces. Hence

$$4\pi = \sum\nolimits_{\text{faces}} [(\text{angle sum}) - (\text{number of sides})\pi + 2\pi].$$

The sum over the faces of the angle sums is $2\pi V$ since the angles surrounding each vertex contribute 2π to the sum. The sum of the number of sides of all the polygons is $2E$ since each edge is shared by two polygons, and thus

$$4\pi = 2\pi V - 2\pi E + 2\pi F = 2\pi(V - E + F),$$

that is, $V - E + F = 2$. ∎

It is easy to project the previous spherical configuration into the plane to obtain a planar graph and to prove Euler's formula by induction on the number of faces, or to draw the graph with line segments to deduce the formula by computing angles in the configuration.

11.9   Faces and vertices in convex polyhedra

What is the domain of Euler's polyhedral formula $V - E + F = 2$? Clearly $V \geq 4$ and $F \geq 4$, consequently $E \geq 6$. But more is true: The average number of sides per face is $2E/F$ (each edge is a side of two faces), and this average is at least 3, hence $2E \geq 3F$. Similarly computing the average number of edges per vertex yields $2E \geq 3V$. These inequalities can rule out particular sets of values of V, E, and F satisfying Euler's formula and the domain restrictions as convex polyhedra. For example, is there a convex polyhedron with seven faces and eleven vertices? If so, then it has sixteen edges, but then $2E = 32 < 33 = 3V$, so the answer is no. See Challenge 11.3 for further inequalities for V, E, and F.

More information about the nature of the faces and vertices can be obtained by introducing the numbers F_n for the number of faces that are n-gons, and V_n for the number of vertices of degree n (the *degree* of a vertex is the number of edges that meet there). Then F and V are the finite sums $\sum_{n \geq 3} F_n$ and $\sum_{n \geq 3} V_n$, respectively. Since each edge is a side of two faces and joins two vertices, we have

$$2E = \sum_{n \geq 3} n F_n = \sum_{n \geq 3} n V_n.$$

From Challenge 11.3b we have $3F - E \geq 6$ or $6F - 2E \geq 12$, and thus

$$6 \sum_{n \geq 3} F_n - \sum_{n \geq 3} n F_n \geq 12,$$

which simplifies to

$$3F_3 + 2F_4 + F_5 \geq 12 + \sum_{n \geq 7} (n - 6) F_n.$$

A completely analogous calculation yields $3V_3 + 2V_4 + V_5 \geq 12 + \sum_{n \geq 7} (n - 6)V_n$, and thus we have proved

Theorem 11.10. *In any convex polyhedron, $3F_3 + 2F_4 + F_5 \geq 12$ and $3V_3 + 2V_4 + V_5 \geq 12$, so that there is always at least one triangle, quadrilateral, or pentagon among the faces and at least one vertex of degree 3, 4, or 5.*

A polyhedron is *regular* if all of its faces are congruent regular polygons and every vertex has the same degree. Convex regular polyhedra are also known as *Platonic solids*. As a consequence of Theorem 11.10, we have

Theorem 11.11. *There are exactly five types of convex regular polyhedra.*

Proof. We use the symbol $\{n, k\}$ to denote a convex polyhedron (if it exists) all of whose faces are n-gons and whose vertices all have degree k. From Theorem 11.10 we know that both n and k are 3, 4, or 5. Since every edge belongs to two faces and joins two vertices, $2E = nF$ and $2E = kV$, so that $F = 2E/n$ and $V = 2E/k$. Substituting these into Euler's formula and solving for E yields $E = 2nk/(2n + 2k - nk)$, and consequently $F = 4k/(2n + 2k - nk)$ and $V = 4n/(2n + 2k - nk)$. Hence $2n + 2k - nk > 0$, or equivalently, $(n - 2)(k - 2) < 4$. Thus the only possible convex polyhedra of the form $\{n, k\}$ are $\{3,3\}$, $\{4,3\}$, $\{3,4\}$, $\{5,3\}$, and $\{3,5\}$. ∎

At this point we only know these are five potential convex polyhedra of the form $\{n, k\}$. We have not assumed that the faces are regular, or even equilateral, equiangular, or congruent. Thus it is rather surprising that all five can be constructed with regular polygonal faces. Table 11.1 gives the relevant facts about the five convex regular polyhedra, and Figure 11.11 illustrates them.

Table 11.1. The five convex regular polyhedra

$\{n, k\}$	E	V	F	common name
$\{3,3\}$	6	4	4	tetrahedron
$\{4,3\}$	12	8	6	cube
$\{3,4\}$	12	6	8	octahedron
$\{5,3\}$	30	20	12	dodecahedron
$\{3,5\}$	30	12	20	icosahedron

Figure 11.11.

Folding and unfolding polyhedra

A common method for constructing a model of a polyhedron is to cut out a plane polygon from a sheet of paper, fold it, and tape edges together to form the polyhedron. For example, the Latin cross in Figure 11.12a can be folded and taped to form a cube. This works because the cube can be unfolded to form the Latin cross (and a variety of other polygons consisting of six squares). An area of current research is to find suitable unfoldings of polyhedra to flat polygons, and the complementary problem of whether or not a given polygon can be folded and taped to form a convex polyhedron [O'Rourke, 2009].

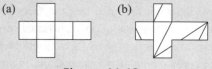

Figure 11.12.

For example, the Latin cross can be folded into at least 23 different convex polyhedra [Demaine and O'Rourke, 2007]. Figure 11.12b illustrates the folds for an irregular tetrahedron.

11.10   Why some types of faces repeat in polyhedra

Every polyhedron must have at least a pair of faces with the same number of sides (see Challenge 11.2). But many polyhedra have three or more faces with the same number of sides. That is true of every polyhedron that we have

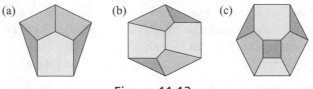

Figure 11.13.

seen so far, but it is not true for every polyhedron. In Figure 11.13 we see (viewed from above) three polyhedra where faces with the same number of sides repeat at most twice.

Again letting F_n denote the number of faces that are n-gons, we have in (a) $F_3 = F_4 = F_5 = 2$, in (b) $F_3 = F_4 = F_5 = 2$ and $F_6 = 1$, and in (c) $F_3 = F_4 = F_5 = F_6 = 2$. We now prove the surprising result that these are the only cases in which a face does not repeat at least three times.

Let $F = F_3 + F_4 + \cdots + F_k$ and assume that $F_i \leq 2$ for all i. Using the inequality $6 \leq 3F - E$ from Challenge 11.3b yields

$$12 \leq 6F - 2E$$
$$= 6\sum_{i=3}^{k} F_i - \sum_{i=3}^{k} iF_i$$
$$= 3F_3 + 2F_4 + F_5 + \sum_{i=7}^{k} (6-i)F_i \leq 3\cdot 2 + 2\cdot 2 + 2 + 0 = 12,$$

whence the inequalities are equalities so that $F_3 = F_4 = F_5 = 2$, F_6 is 0, 1, or 2, and $F_i = 0$ for $i \geq 7$.

11.11 Euler and Descartes à la Pólya

Given a convex polyhedron with V vertices $v_1, v_2, \ldots, v_V$, consider the *angular defect* Δ_i of vertex v_i, defined as the difference between 2π and the sum of the plane angles around v_i. Let Δ denote the total angular defect for the polyhedron, i.e., $\Delta = \Delta_1 + \Delta_2 + \cdots + \Delta_V$. For example, for the cube $\Delta_i = \pi/2$ at each of the eight vertices (see Figure 11.14a) so that $\Delta = 4\pi$; and for the icosahedron, $\Delta_i = \pi/3$ at each of the twelve vertices (see Figure 11.14b) so that again $\Delta = 4\pi$.

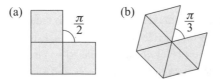

Figure 11.14.

René Descartes (1596–1650) discovered that $\Delta = 4\pi$ holds for every convex polyhedron, a result now known as *Descartes' Angular Defect Theorem*. The relationship between Descartes' theorem and Euler's polyhedral formula is given in the next theorem. Our proof is from [Pólya, 1965].

Descartes' Angular Defect Theorem 11.12. $\Delta = 2\pi(V - E + F)$.

Proof. We let S denote the sum of all the angles in all the faces of the polygon, and evaluate S in two ways. The sum of the angles at vertex v_i is $2\pi - \Delta_i$ so that computing S vertex-by-vertex yields

$$S = (2\pi - \Delta_1) + (2\pi - \Delta_2) + \cdots + (2\pi - \Delta_V) = 2\pi V - \Delta.$$

Again let F_n denote the number of faces that are n-gons, so that $F = \sum_{n \geq 3} F_n$. Since the angle sum of an n-gon is $(n-2)\pi$, computing S face-by-face yields

$$S = \sum_{n \geq 3} (n-2)\pi F_n = \pi \sum_{n \geq 3} n F_n - 2\pi \sum_{n \geq 3} F_n = (2E - 2F)\pi.$$

Thus $2\pi V - \Delta = (2E - 2F)\pi$, or $\Delta = 2\pi(V - E + F)$ as claimed. ∎

As a consequence, $\Delta = 4\pi$ if and only if $V - E + F = 2$, establishing the logical equivalence of Descartes' theorem and Euler's formula.

Was Euler aware of Descartes' theorem?

Descartes' theorem appears in his work *Progymnasmata de Solidorum Elementis* (*Exercises on the Elements of Solids*), a work that was not published in his lifetime and, indeed, was lost until 1860 when a copy was discovered in the papers of Gottfried Wilhelm Leibniz [Cromwell, 1997]. Since Euler was born 57 years after Descartes' death, and died 77 years before the discovery of the *Progymnasmata*, he could not possibly have known of Descartes' result.

11.12   Squaring squares and cubing cubes

We *square* a square or rectangle by tiling it with smaller squares. A squared square or rectangle is *simple* if it does not contain a smaller squared square or rectangle, and *compound* if it does. A squared square or rectangle is *perfect* if all the square tiles are of different sizes, and *imperfect* if they are not. The *order* of a squared square or rectangle is the number of square tiles it

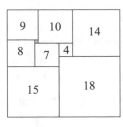

Figure 11.15.

contains. Figure 11.15 illustrates an order 9 simple perfect squared 32×33 rectangle. The small gray square has side length 1, and the others have the indicated side length.

It is fairly easy to create squared rectangles; squared squares are more difficult [Gardner, 1961; Honsberger, 1970]. The smallest known simple perfect squared square is an order 21 square with side length 112.

Can we cube a cube or rectangular box in an analogous fashion?

Theorem 11.13. *No cubed rectangular boxes of finite order exist.*

Proof [Gardner, 1961]. Suppose a cubed box exists, and is sitting before you on the table. The bottom face of the box is a squared rectangle, which will contain a smallest square. Since this square cannot be on the edge of the rectangular base, it must be the bottom face of a cube (which we denote as cube A) that is completely surrounded by larger cubes (see Figure 11.16).

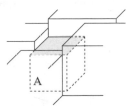

Figure 11.16.

Still smaller cubes sit on the top of cube A, forming a squared square on its top face. In this squared square will be a smallest square, the bottom face of cube B, the smallest cube resting on the top of cube A. Continuing, we need a smallest cube, cube C, resting on the top face of cube B. The argument continues, calling for an infinite collection of smaller and smaller cubes in the box. Thus no rectangular box can be partitioned into a finite number of cubes of different sizes. ∎

11.13   Challenges

11.1 The reciprocal Pythagorean theorem (Theorem 5.1) can be extended to right tetrahedra. In Figure 11.1 let p denote the altitude to the base $\triangle ABC$. Show that $(1/p)^2 = (1/a)^2 + (1/b)^2 + (1/c)^2$.

11.2 Prove that every polyhedron must have at least two faces with the same number of sides.

11.3 Let V, E, and F denote the number of vertices, edges, and faces, respectively, in a convex polyhedron. Prove that (a) $2V - F \geq 4$ and $2F - V \geq 4$, and (b) $3F - E \geq 6$ and $3V - E \geq 6$.

11.4 In a letter to Christian Goldbach (1690–1764) Euler wrote that no polyhedron has exactly seven edges. Prove that Euler was correct (as usual) [Cromwell, 1997].

11.5 Prove that the number of different ways to color a Platonic solid (with F faces and E edges) with F colors so that each face has a different color is $F!/(2E)$. Two colorings of the solid are the same if one of them can be rotated in space so that it appears identical to the other with corresponding faces colored alike.

11.6 Let P be a square pyramid (four equilateral triangles on a square base) and T a regular tetrahedron with faces congruent to the triangular faces of P. If P and T are joined by gluing one of the faces of T onto one of the faces of P, how many faces does the resulting polyhedron have? (Hint: it is not 7!)

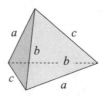

Figure 11.17.

11.7 An *isosceles tetrahedron* is one whose opposite edges are equal in length (see Figure 11.17). Prove that the volume of an isosceles tetrahedron with edge lengths a, b, and c is given by

$$\sqrt{(a^2 + b^2 - c^2)(c^2 + a^2 - b^2)(b^2 + c^2 - a^2)/72},$$

and that the diameter of the circumscribed sphere is $\sqrt{(a^2 + b^2 + c^2)/2}$.

CHAPTER 12

Additional Theorems, Problems, and Proofs

In our final chapter, we present a collection of theorems and problems from various branches of mathematics and their proofs and solutions. We begin by discussing some set theoretic results concerning infinite sets, including the Cantor-Schröder-Bernstein theorem. In the next two sections we present proofs of the Cauchy-Schwarz inequality and the AM-GM inequality for sets of size n. We then use origami to solve the classical problems of trisecting angles and doubling cubes, followed by a proof that the Peaucellier-Lipkin linkage draws a straight line. We then look at several gems from the theory of functional equations and inequalities. In the final sections we conclude with an infinite series and an infinite product for simple expressions involving π, and illustrate each with an application.

12.1 Denumerable and nondenumerble sets

> *The infinite! No other question has ever moved so profoundly the spirit of man; no other idea has so fruitfully stimulated his intellect; yet no other concept stands in greater need of clarification than that of the infinite.*
>
> David Hilbert

> *To infinity, and beyond!*
>
> Buzz Lightyear, *Toy Story* (1995)

Two sets *have the same cardinality* if there exists a one-to-one function from one set onto the other, i.e., a one-to-one correspondence between the sets. An infinite set of numbers is *denumerable* (or *countably infinite*) if it has the same cardinality as the set $\mathbb{N} = \{1, 2, 3, \dots\}$ of natural (or counting) numbers. For example, the set $\mathbb{Z} = \{\dots, -3, -2, -1, 0, 1, 2, 3, \dots\}$ of integers

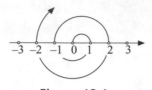

Figure 12.1.

is denumerable, since the function $f : \mathbb{N} \to \mathbb{Z}$ defined by

$$f(n) = \begin{cases} n/2, & n \text{ even,} \\ (1-n)/2, & n \text{ odd,} \end{cases}$$

is clearly one-to-one and onto. This one-to-one correspondence can be illustrated graphically with an Archimedean-like spiral (see Section 9.1) as shown in Figure 12.1.

The set $\mathbb{Z} \times \mathbb{Z}$ of ordered pairs of integers is also denumerable, and the one-to-one correspondence is nicely illustrated [MacHale, 2004] in Figure 12.2.

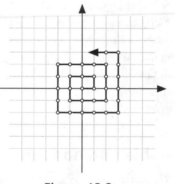

Figure 12.2.

Almost everyone knows that the set $\mathbb{Q}$ of rational numbers is denumerable. There are many proofs of this fact, but most suffer from the defect that an explicit one-to-one correspondence between $\mathbb{Q}$ and $\mathbb{N}$ is not exhibited. Many of the proofs construct two one-to-one functions, one from each set into the other. This approach then requires the Cantor-Schröder-Bernstein theorem (see the next section) to conclude the two sets have the same cardinality.

To show that $\mathbb{Q}$ is denumerable, it suffices to show that the set $\mathbb{Q}_+$ of positive rationals is denumerable (see Challenge 12.1), which we do with the following one-to-one correspondence between $\mathbb{Q}_+$ and $\mathbb{N}$ [Sagher, 1989]. Let a/b be a positive rational number with a and b relatively prime. Let

$a = p_1^{e_1} p_2^{e_2} \cdots p_m^{e_m}$ and $b = q_1^{f_1} q_2^{f_2} \cdots q_m^{f_m}$ be the prime number decompositions of a and b. Define a function $g : \mathbb{Q}_+ \to \mathbb{N}$ by $g(1) = 1$ and

$$g(a/b) = p_1^{2e_1} p_2^{2e_2} \cdots p_m^{2e_m} q_1^{2f_1-1} q_2^{2f_2-1} \cdots q_m^{2f_m-1}.$$

The function g is clearly one-to-one and onto, hence $\mathbb{Q}_+$ (and thus $\mathbb{Q}$) is denumerable.

12.2 The Cantor-Schröder-Bernstein theorem

The Cantor-Schröder-Bernstein theorem is a staple of set theory. It is used to show that two infinite sets have the same cardinality without finding an explicit one-to-one correspondence between them. It is named after Georg Cantor (1845–1918) who conjectured but did not prove it, Ernst Schröder (1841–1902) who published the first proof (later found to be incorrect), and Felix Bernstein (1878–1956), who gave the first correct proof.

The Cantor-Schröder-Bernstein Theorem 12.1. *If X and Y are given nonempty sets, f a one-to-one function from X onto a subset of Y, and g a one-to-one function from Y onto a subset of X, then there exists a one-to-correspondence between X and Y.*

Proof. [Schweizer, 2000]. Without loss of generality we can assume that X and Y are disjoint (if they are not, replace X and Y with $X \times \{a\}$ and $Y \times \{b\}$, respectively, where a and b are any two distinct objects not elements of $X \cup Y$). For any x in X, define the *orbit* of x to be the set

$$\left\{ \ldots, g^{-1} f^{-1} g^{-1}(x), f^{-1} g^{-1}(x), g^{-1}(x), x, f(x), gf(x), fgf(x), \ldots \right\},$$

where f^{-1} and g^{-1} are the inverses of f and g, respectively, and juxtaposition denotes composition of functions. We can illustrate this set with a directed graph in which, for any x in X and y in Y, there is a directed edge from x to y (denoted $x \to y$) if $y = f(x)$ and a directed edge from y to x (denoted $y \to x$) if $x = g(y)$. Since f and g are one-to-one functions, two orbits are either distinct or identical, i.e., the set of all orbits is a partition of $X \cup Y$. As a consequence, each orbit falls into precisely one of four distinct classes, which are illustrated in Figure 12.3, where the xs and ys denote distinct elements of X and Y.

Since every element of $X \cup Y$ belongs to a unique orbit of one of the four types in Figure 12.3, the desired one-to-one correspondence $\phi : X \to Y$ can be defined as follows: (a) if x belongs to an orbit of type I, III, or IV,

$$\text{I.}\quad x \rightarrow y \rightarrow x \rightarrow y \rightarrow \cdots$$
$$\text{II.}\quad y \rightarrow x \rightarrow y \rightarrow x \rightarrow \cdots$$
$$\text{III.}\quad \cdots \rightarrow x \rightarrow y \rightarrow x \rightarrow y \rightarrow \cdots$$

$$
\begin{array}{ccccccc}
& x \rightarrow y \rightarrow x \rightarrow y \rightarrow x \rightarrow y \\
& \uparrow & & & & \downarrow \\
\text{IV.} \quad y & & & & & x \\
& \uparrow & & & & \downarrow \\
& x \leftarrow y \leftarrow \cdots \quad \cdots \leftarrow x \leftarrow y
\end{array}
$$

Figure 12.3.

let $\phi(x) = f(x)$; and (b) if x belongs to an orbit of type II, let $\phi(x) = g^{-1}(x)$. The function ϕ is clearly one-to-one and onto, which completes the proof. ∎

As an example, we give another simple proof [Campbell, 1986] that $\mathbb{Q}$ is denumerable by applying the Cantor-Schröder-Bernstein theorem to $\mathbb{Q}_+$ and $\mathbb{N}$. Clearly there exists a one-to-one function (the identity) from $\mathbb{N}$ onto a subset of $\mathbb{Q}_+$. In the other direction, note that each symbol a/b for a positive rational is a distinct positive integer in base 11 with / as the symbol for 10, which completes the proof. For example,

$$22/7 = 2(11^3) + 2(11^2) + 10(11) + 7 = 3021.$$

Since no assumption that $(a, b) = 1$ has been made, this proof actually shows that the set of all representations of rationals is denumerable.

12.3   The Cauchy-Schwarz inequality

In 1821 Augustin-Louis Cauchy (1789–1857) published the following inequality that now bears his name:

Theorem 12.2. *For any two sets $\{a_1, a_2, \ldots, a_n\}$ and $\{b_1, b_2, \ldots, b_n\}$ of real numbers, we have*

$$\left| \sum\nolimits_{k=1}^{n} a_k b_k \right| \le \sqrt{\sum\nolimits_{k=1}^{n} a_k^2} \sqrt{\sum\nolimits_{k=1}^{n} b_k^2}, \tag{12.1}$$

with equality if and only if the two sets are proportional, i.e. $a_i b_j = a_j b_i$ for all i and j between 1 and n.

Proof. Integral versions of this inequality were published by Victor Yacovlevich Bunyakovski (1804–1889) in 1859 and Hermann Amandus Schwarz

(1843–1921) in 1885. Today the inequality is known either as the Cauchy-Schwarz inequality or the Cauchy-Bunyakovski-Schwarz inequality.

As promised in Section 7.3, here is an elegant one-line proof using only the triangle inequality and the AM-GM inequality for two numbers. For non-negative x and y, $xy \leq (x^2 + y^2)/2$ and hence (all the sums are for k from 1 to n)

$$\frac{|\sum a_k b_k|}{\sqrt{\sum a_k^2}\sqrt{\sum b_k^2}} \leq \sum \frac{|a_k|}{\sqrt{\sum a_k^2}} \cdot \frac{|b_k|}{\sqrt{\sum b_k^2}} \leq \frac{1}{2} \sum \left(\frac{|a_k|^2}{\sum a_k^2} + \frac{|b_k|^2}{\sum b_k^2} \right) = 1.$$

∎

The inequality and the proof also apply to infinite sequences that are *square-summable*, i.e., sequences $\{a_k\}_{k=1}^{\infty}$ such that $\sum_{k=1}^{\infty} a_k^2$ is finite.

In the other direction, the Cauchy-Schwarz inequality for n numbers also implies the AM-GM inequality for two numbers; see Challenge 12.4.

There are many other proofs of the Cauchy-Schwarz inequality. For a collection, many of them visual, see [Alsina and Nelsen, 2009].

It is rather amusing that the Cauchy-Schwarz inequality can be used to establish an inequality between the golden ratio φ (see Section 1.6) and the sum $s = \pi^2/6$ of the series $1 + 1/4 + 1/9 + \cdots + 1/n^2 + \cdots$ (see Section 12.9).

Corollary 12.1. $\pi^2/6 > \varphi$.

Proof. Consider two sequences, $\{a_k\}_{k=1}^{\infty}$ and $\{b_k\}_{k=1}^{\infty}$, with $a_k = 1/k$ and $b_k = 1/(k + 1)$. Both sequences are square-summable, with $\sum_{k=1}^{\infty} a_k^2 = s$ and $\sum_{k=1}^{\infty} b_k^2 = s - 1$, where $s = \pi^2/6$. The Cauchy-Schwarz inequality yields

$$\left(\sum_{k=1}^{\infty} 1/[k(k + 1)] \right)^2 = \left(\sum_{k=1}^{\infty} a_k b_k \right)^2 < \sum_{k=1}^{\infty} a_k^2 \sum_{k=1}^{\infty} b_k^2 = s(s-1).$$

However, the series in the parentheses on the left telescopes to 1, and thus $s^2 - s - 1 > 0$. Since φ is the positive root of $x^2 - x - 1 = 0$ and $s^2 - s - 1 > 0$ with s positive, s must be larger than φ, i.e., $\pi^2/6 > \varphi$. See Figure 12.4. ∎

The two numbers φ and $\pi^2/6$ are actually quite close: to five decimal places, we have $\varphi \approx 1.61803 < 1.64493 \approx \pi^2/6$.

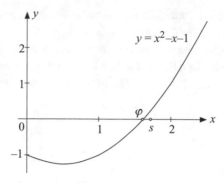

Figure 12.4.

12.4 The arithmetic mean-geometric mean inequality

The AM-GM (arithmetic mean-geometric mean) inequality for n positive numbers $a_1, a_2, \cdots, a_n$ states that

$$\frac{a_1 + a_2 + \cdots + a_n}{n} \geq (a_1 a_2 \cdots a_n)^{1/n} \qquad (12.2)$$

with equality if and only if $a_1 = a_2 = \cdots = a_n$. This inequality is "the fundamental theorem of the theory of inequalities, the keystone on which many other very important results rest," and "one that can be established in a large number of interesting ways; there are literally dozens of different proofs based on ideas from a great variety of sources" [Beckenbach and Bellman, 1961]. Indeed, one can find over fifty proofs in [Bullen et al., 1988].

For $n = 2$ the inequality is $(a + b)/2 \geq \sqrt{ab}$ for positive numbers a and b, and is readily proved algebraically by expanding and simplifying $(\sqrt{a} - \sqrt{b})^2 \geq 0$. There is also a variety of geometric proofs, for example the simple one is exhibited in Figure 12.5. The sum of the areas $a/2$ and $b/2$ of

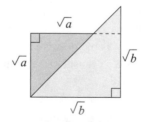

Figure 12.5.

the shaded triangles is at least as great as the area $\sqrt{ab}$ of the rectangle with base $\sqrt{b}$ and height $\sqrt{a}$.

Many of the proofs of (12.2) use mathematical induction. Induction proofs are rarely charming so here is a direct proof. To prove (12.2) we prove a more general version, sometimes referred to as *the generalized AM-GM inequality* for weighted arithmetic and geometric means.

Theorem 12.3. *For positive real numbers $a_1, a_2, \ldots, a_n$ and positive real numbers $\lambda_1, \lambda_2, \ldots, \lambda_n$ that sum to one, we have*

$$\lambda_1 a_1 + \lambda_2 a_2 + \cdots + \lambda_n a_n \geq a_1^{\lambda_1} a_2^{\lambda_2} \cdots a_n^{\lambda_n} \qquad (12.3)$$

with equality if and only if $a_1 = a_2 = \cdots = a_n$.

Our proof is from [Steele, 2004], where it is attributed to George Pólya. The proof uses the solution to Steiner's problem on the number e from Section 2.10: $e^{1/e} \geq x^{1/x}$ for $x > 0$, with equality if and only if $x = e$. Inequality (12.2) is the special case of (12.3) where $\lambda_1 = \lambda_2 = \cdots = \lambda_n = 1/n$.

Proof. The inequality from Steiner's problem is equivalent to $e^{x/e} \geq x$. Now let $y = x/e$, thus $e^y \geq ey$ for $y > 0$ with equality if and only if $y = 1$. Let A equal the weighted arithmetic mean $\lambda_1 a_1 + \lambda_2 a_2 + \cdots + \lambda_n a_n$, let G equal the weighted geometric mean $a_1^{\lambda_1} a_2^{\lambda_2} \cdots a_n^{\lambda_n}$, and set $y = a_i/A$:

$$e^{a_i/A} \geq ea_i/A.$$

Now raise each side to the λ_i power:

$$e^{\lambda_i a_i/A} \geq (ea_i/A)^{\lambda_i} = (e/A)^{\lambda_i} a_i^{\lambda_i}.$$

Multiplying the inequalities for $i = 1, 2, \ldots, n$ and recalling that $\lambda_1 + \lambda_2 + \cdots + \lambda_n = 1$ yields

$$e = \exp\left(\frac{\lambda_1 a_1 + \lambda_2 a_2 + \cdots + \lambda_n a_n}{A}\right)$$

$$\geq \left(\frac{e}{A}\right)^{\lambda_1 + \lambda_2 + \cdots + \lambda_n} a_1^{\lambda_1} a_2^{\lambda_2} \cdots a_n^{\lambda_n} = \frac{eG}{A},$$

and thus $A \geq G$. The inequality is strict unless $a_1 = a_2 = \cdots = a_n = A$, which completes the proof. ∎

Since the generalized AM-GM inequality can be used to prove $e^{1/e} \geq x^{1/x}$ for $x > 0$, the two inequalities are equivalent. See Challenge 12.5.

12.5 Two pearls of origami

The three classical problems of antiquity are *doubling the cube* (construct a cube whose volume is twice that of a given cube), *trisecting the angle* (trisect any given angle), and *squaring the circle* (construct a square whose area equals that of a given circle). Although it is impossible to solve them with the tools used by the ancient Greeks—straightedge and compass—they can be solved with more sophisticated tools. For example, in Section 9.2 we used Archimedean spirals to trisect angles and square circles, and in two of the Chapter 9 Challenges we used other curves to double cubes.

In this section we will show haw to solve two of these problems using another sophisticated tool, one we encountered in Section 4.3—origami. But first, what sort of geometrical operations can be accomplished with origami? Some mathematical principles of origami are captured by the following seven axioms (the first six are due to Humiaki Huzita [Huzita, 1992], the seventh is due to Koshiro Hatori in 2002 [Hatori, 2009]). Our presentation and illustration of the axioms is from [Lang, 2003].

The Huzita-Hatori origami axioms

O1 Given two points p_1 and p_2, there is a fold passing through both of them.

O2 Given two points p_1 and p_2, there is a fold that places p_1 onto p_2.

O3 Given two lines ℓ_1 and ℓ_2, there is a fold that places ℓ_1 onto ℓ_2.

O4 Given a point p and a line ℓ, there is a fold perpendicular to ℓ that passes through p.

O5 Given two points p_1 and p_2 and a line ℓ, there is a fold that places p_1 onto ℓ and passes through p_2.

O6 Given two points p_1 and p_2 and two lines ℓ_1 and ℓ_2, there is a fold that places p_1 onto ℓ_1 and place p_2 onto ℓ_2.

O7 Given a point p and two lines ℓ_1 and ℓ_2, there is a fold perpendicular to ℓ_2 that places p onto ℓ_1.

See Figure 12.6 for an illustration of the seven axioms. It is worth noting that the axioms refer to the *existence* of certain folds and not to the instructions for actually making them, some of which can be rather complex.

To illustrate the power of origami folds, we now solve two of the classical problems mentioned at the beginning of this section. We first present

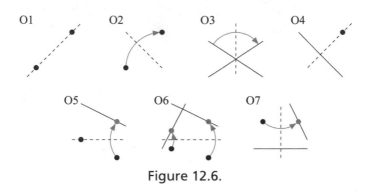

Figure 12.6.

an origami-based method for trisecting an angle due to Tsune Abe [Fusimi, 1980].

If we can trisect acute angles we can trisect all angles, since it is easy to trisect an integer multiple of $90°$. On a sheet of paper place the acute angle θ to be trisected so that the vertex is in the lower left corner, one side the bottom edge of the paper, and the other side line ℓ_2, as illustrated in Figure 12.7a. Make a horizontal fold about half way between the top and the bottom, then make an Axiom O3 fold placing the bottom of the paper on that fold, producing line ℓ_1, equidistant between the bottom and the first fold. Now perform an Axiom O6 fold, placing p_1 onto ℓ_1 and p_2 onto ℓ_2, as illustrated in Figure 12.7b.

With the paper folded as in Figure 12.7b, re-fold along the ℓ_1 crease to obtain the fold labeled ℓ_3. Now unfold the paper, and fold along the ℓ_3 crease. This fold will pass through the lower left corner, and the angle between ℓ_2 and ℓ_3 will be $\theta/3$.

Figure 12.7c shows why Abe's method trisects the angle. Since $|OA| = |OC|$, $|AB| = |BC| = |CD|$, and $OB \perp AC$, OB will pass through vertex

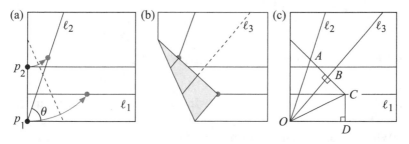

Figure 12.7.

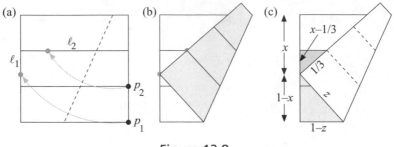

Figure 12.8.

O of isosceles triangle OAC, right triangles OAB, OBC, and OCD are congruent, and thus $\angle OAB = \angle OBC = \angle OCD = \theta/3$.

The second classical problem we solve with origami is doubling a cube. To do so, we need to construct a line segment with length $a = \sqrt[3]{2}$, since a cube with this edge length has volume 2, twice the volume of a cube with edge length 1. We begin with a square piece of paper with edge length 1 that has been divided into thirds, as illustrated in Figure 12.8a, and again perform an Axiom O6 fold to place p_1 onto ℓ_1 and p_2 onto ℓ_2, as shown in Figure 12.8b. We claim that the image of p_1 partitions the left side of the square into two segments whose ratio is $\sqrt[3]{2}$.

To prove this, we need to show that $x/(1-x) = \sqrt[3]{2}$ in Figure 12.8c. The two shaded right triangles are similar, and hence

$$3x - 1 = \frac{x - 1/3}{1/3} = \frac{1-z}{z} = \frac{1}{z} - 1,$$

or, equivalently, $3xz = 1$, or $3x(2z) = 2$. But in the light gray triangle we have $(1 - x)^2 + (1 - z)^2 = z^2$, or $2z = (1 - x)^2 + 1$. Hence $3x[(1 - x)^2 + 1] = 2$, which is equivalent to $x^3 = 2(1 - x)^3$, and hence $x/(1 - x) = \sqrt[3]{2}$, as claimed. So $\sqrt[3]{2} = 1/(1 - x) - 1$, and this length can be easily marked on paper once a segment of length $1 - x$ has been obtained.

12.6 How to draw a straight line

It is easy to do the construction in the title: Use a straightedge. But how do we know that the straightedge is straight? Indeed, do we know what "straight" really means?

When we draw a circle, we usually do not trace a circular disk with a pencil or pen, we use (or used to use, before computer software existed) a

Figure 12.9.

compass. The compass is a mechanical device that enables us to implement the definition of the circle as the locus of points equidistant from a fixed point. Euclid's definition of a straight line as "a line (which he defines as 'breadthless length') which lies evenly with the points on itself" (Definition 4 in Book I of the *Elements*) does not seem to help much with actually drawing a line.

Is it possible to create a mechanical device for drawing a straight line, analogous to the compass for drawing circles? This was an important question in the 19th century, when many mechanical devices were invented, motivated by the industrial revolution. Linkages (rigid bars made of metal or wood joined by rivets at their ends) were created to turn circular motion into linear motion, and vice-versa. The Scottish engineer James Watt (1736–1819) and the Russian mathematician Pafnuty Lvovich Chebyshev (1821–1894) created linkages to produce approximately linear motion, but the first to create a linkage to produce truly linear motion from circular motion was one designed by the French engineer Charles-Nicolas Peaucellier (1832–1913) in 1864, and independently rediscovered by the Russian mathematician Lippman Lipkin (1851–1875) in 1871. This linkage is now variously known as the *Peaucellier-Lipkin linkage*, the *Peaucellier cell*, or the *Peaucellier inversor*.

In 1876 Alfred Bray Kempe (whom we met in Chapter 10 as the author of a faulty proof of the Four Color Theorem) gave a lecture at London's South Kensington Museum with the title of this section, which was published as a small book the following year [Kempe, 1877]. In Figure 12.9 we see Kempe's illustration of the Peaucellier-Lipkin linkage.

In his lecture and book, Kempe presented a sophisticated argument showing that the linkage describes a straight line. Our argument is simpler, based on the law of cosines. In Figure 12.10, the lengths of the parts of the linkage are $|BC| = |BD| = a$, $|AC| = |CP| = |AD| = |DP| = b$ with $a > b > 0$, and $|AE| = |BE| = r$. Since B and E are fixed points, point A will be on a circle of radius r centered at E. Let $\angle ABE = \alpha$, $\angle CAP = \beta$, and let Q be the foot of the perpendicular from P to the x-axis.

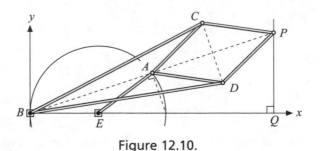

Figure 12.10.

To show that P describes a straight line, we show that the x-coordinate of P depends only on a, b, and r, and not on α or β. Since $|AB| = 2r \cos \alpha$ and $|AP| = 2b \cos \beta$, the x-coordinate of P is $|BQ| = |BP| \cos \alpha = (2r \cos \alpha + 2b \cos \beta) \cos \alpha$. Applying the law of cosines to triangle ABC yields

$$a^2 = b^2 + (2r \cos \alpha)^2 - 2b(2r \cos \alpha) \cos(\pi - \beta),$$

whence

$$\frac{a^2 - b^2}{2r} = (2r \cos \alpha + 2b \cos \beta) \cos \alpha = |BQ|.$$

Thus the x-coordinate of P depends only on a, b, and r, so that P describes a straight line as claimed.

12.7   Some gems in functional equations

In high school algebra courses we learn how to find solutions to algebraic equations whose unknowns are real numbers. In the first course in calculus we learn how to find solutions to simple differential equations whose unknowns are functions. Functional equations are like differential equations in that the unknowns are functions, but the equations usually do not involve derivatives or integrals. Functional equations often appear in mathematical competitions such as national and international Mathematical Olympiads and the William Lowell Putnam Competition. Functional equation problems are often the most difficult competition problems to solve since their solutions often require only some elementary mathematics—but much ingenuity.

> **János Aczél and the theory of functional equations**
>
> Functional equations have a long and distinguished mathematical history. Many mathematicians have done seminal work in the field: Nicole Oresme (1323–1382), Jean le Rond d'Alembert (1717–1783), Augustin-Louis Cauchy (1789–1857), Leonhard Euler (1707–1783), Neils Henrik Abel (1802–1829), David Hilbert (1862–1943), and others. But during much of this time, functional equations were solved primarily by ad hoc methods and ingenuity. The first systematic treatment—classifying the equations and their methods of solution—was János Aczél's *Vorlesungen über Funktionalgleichungen und ihre Anwendungen* (*Lectures on Functional Equations and their Applications*) in 1961 and its subsequent English translation [Aczél, 1966]. It is still a key reference in the field.

A function $f : \mathbb{N} \to \mathbb{Z}$ is called *completely multiplicative* if $f(1) = 1$ and if f satisfies the functional equation $f(mn) = f(m)f(n)$ for all positive integers m and n. Examples of completely multiplicative functions are $g(x) = x^k$ for some k in $\mathbb{N}$ and $h(x) = 1$ for $x = 1$, $h(x) = 0$ for $x > 1$.

In the next theorem we show that imposing additional requirements can lead to a unique solution to a functional equation.

Theorem 12.4. *Let $f : \mathbb{N} \to \mathbb{Z}$ be a strictly increasing completely multiplicative function with $f(2) = 2$. Then $f(n) = n$ for all n in $\mathbb{N}$.*

Proof. For any $k \geq 0$ induction establishes $f(2^k) = [f(2)]^k = 2^k$. Since f is strictly increasing, we have

$$2^k = f(2^k) < f(2^k + 1) < f(2^k + 2) < \cdots < f(2^{k+1} - 1)$$
$$< f(2^{k+1}) = 2^{k+1}.$$

But there are only $2^{k+1} - 2^k - 1 = 2^k - 1$ different integers strictly between 2^k and 2^{k+1}, and thus $f(2^k + j) = 2^k + j$ for j between 1 and $2^k - 1$ inclusive, thus $f(n) = n$ for all n in $\mathbb{N}$. ∎

In the next theorem and proof, the solution to the functional equation is a property possessed by the function rather than an explicit formula.

Theorem 12.5. *Let c be a fixed nonzero real number, and let $f : \mathbb{R} \to \mathbb{R}\backslash\{0\}$ such that f satisfies the functional equation $f(x) = f(x - c)f(x + c)$ for all x in $\mathbb{R}$. Then f is a periodic function.*

Proof. Since $f(x + c) = f(x)f(x + 2c) = f(x)f(x + c)f(x + 3c)$ and $f(x + c)$ is nonzero, we have $1 = f(x)f(x + 3c)$ and hence

$$f(x) = \frac{1}{f(x + 3c)} = f(x + 6c).$$

∎

For real values of x and y, the functional equation

$$f(x + y) = f(x) + f(y) \tag{12.4}$$

is known as *Cauchy's functional equation*, and it is a cornerstone of the general theory of functional equations. While it is easy to show that the functions $f(x) = kx$ are solutions for an arbitrary real k, we ask: are these are the *only* solutions to (12.4)? Answering this simple sounding question is not trivial. Hence we begin with a simpler problem, finding solutions to Cauchy's functional equation for *rational* x and y.

Theorem 12.6. *Let* $f : \mathbb{Q} \to \mathbb{R}$ *be a function satisfying the functional equation* (12.4). *Then* $f(x) = kx$ *for some real constant k.*

Proof. Setting $x = y = 0$ in (12.4) yields $f(0) = 0$ and setting $y = -x$ yields $0 = f(0) = f(x-x) = f(x) + f(-x)$, hence $f(-x) = -f(x)$. Thus f is an odd function and we need only consider positive values of x and y. Repeated use of (12.4) yields, for any real x and any positive integer n,

$$f(nx) = f(x + x + \cdots + x) = nf(x).$$

Now let $x = (m/n)y$ for positive integers m and n. Then $nx = my$, so that $f(nx) = f(my)$ and hence $nf((m/n)y) = nf(x) = mf(y)$, i.e., $f((m/n)y) = (m/n)f(y)$. Thus for rational q and real y, $f(qy) = qf(y)$, and in particular $f(q) = qf(1)$ for any rational q (here $k = f(1)$). ∎

Now consider a function $f : \mathbb{R} \to \mathbb{R}$ satisfying the functional equation (12.4). By Theorem 12.5, all the values of $f(x)$ for x rational lie on a line. If the complete graph is not a line, then we have the surprising result that the graph must be *dense in the plane*. The graph G of f is the set $G = \{(x, y) \mid y = f(x), \ x \in \mathbb{R}\}$, and a set A is *dense* in a set B if given any y element of B we can always a find an element x of A as close to y as we wish. For example, $\mathbb{Q}$ is dense in $\mathbb{R}$, and the set of ordered pairs of rational numbers is dense in the plane.

Corollary 12.2. *Let* $f : \mathbb{R} \to \mathbb{R}$ *satisfy the functional equation* (12.4), *and suppose that the graph of* $y = f(x)$ *is not a line. Then the graph of f is dense in the plane.*

Proof [Aczél and Dhombres, 1989]. Let a be a nonzero rational number so that $f(a) = ka$, and choose an irrational number b such that $f(b) \neq kb$, or equivalently, $bf(a) - af(b) \neq 0$. Thus for any point (x_0, y_0) in the plane, there exist real numbers ρ and σ such that $\rho(a, f(a)) + \sigma(b, f(b)) = (x_0, y_0)$. Explicitly, $\rho = (y_0 b - x_0 f(b))/(bf(a) - af(b))$ and $\sigma = (x_0 f(a) - y_0 a)/(bf(a) - af(b))$. Hence there exist rational numbers r and s such that $r(a, f(a)) + s(b, f(b))$ is as close to (x_0, y_0) as we wish. However, since r and s are rational and f satisfies (12.4),

$$
\begin{aligned}
r(a, f(a)) + s(b, f(b)) &= (ra + sb, rf(a) + sf(b)) \\
&= (ra + sb, f(ra) + f(sb)) \\
&= (ra + sb, f(ra + sb)).
\end{aligned}
$$

Hence the set $G_0 = \{(x, y) \mid y = f(x), x = ra + sb, r, s \in \mathbb{Q}\}$ is dense in the plane, and since G_0 is a subset of G, G must be dense in the plane as well. ■

If $f : \mathbb{R} \to \mathbb{R}$ satisfies (12.4) and also some property that insures that the graph is *not* dense in the plane, then we can conclude that $f(x) = kx$ for all real x.

Corollary 12.3. *Let $f : \mathbb{R} \to \mathbb{R}$ satisfy the functional equation (12.4) and in addition one of the following properties:*

(i) *f is continuous at some real number;*

(ii) *f is bounded on some interval $[a, b]$ with $a < b$; or*

(iii) *f is monotone on some interval $[a, b]$ with $a < b$.*

Then $f(x) = kx$ for some real constant k.

Cauchy's functional equation (and its solution) can be used to show that certain properties possessed by exponential and logarithmic functions are uniquely possessed by those functions. If b is a positive real number and if x and y are any real numbers, then $b^{x+y} = b^x b^y$, or equivalently, the functions $f(x) = b^x$ are solutions to the functional equation

$$f(x + y) = f(x)f(y). \tag{12.5}$$

Are there any other continuous functions $f : \mathbb{R} \to \mathbb{R}$ that satisfy (12.5)? Clearly $f(x) \equiv 0$ satisfies the equation, so we seek solutions that are not identically zero.

Assume $f : \mathbb{R} \to \mathbb{R}$ is a continuous solution to (12.5) not identically zero. Then $f(0) = f(0 + 0) = [f(0)]^2$, so $f(0) = 0$ or 1. If $f(0) = 0$ then for any x, $f(x) = f(x + 0) = f(x)f(0) = 0$, a contradiction, so $f(0) = 1$. Since $f(x) = [f(x/2)]^2$, we have $f(x) \geq 0$ for all x. Is 0 in the range of f? If so, then for some a, $f(a) = 0$. But then for any x, $f(x) = f(x - a + a) = f(x - a)f(a) = 0$, again a contradiction. Hence $f(x) > 0$ for all x. Thus we can apply natural logarithms to (12.5) to obtain $\ln f(x + y) = \ln f(x) + \ln f(y)$, that is, $\ln f(x)$ is a continuous solution to Cauchy's functional equation. Thus $\ln f(x) = kx$ for some constant k, and $f(x) = e^{kx} = b^x$ where $b = e^k > 0$. Hence the functional equation (12.5) characterizes exponential functions, as they are its only (nontrivial) solutions.

Similarly, the logarithm of a product of two positive numbers is the sum of the logarithms of the numbers, so we ask: What continuous functions $f : (0, \infty) \to \mathbb{R}$ (other than $f(x) \equiv 0$) satisfy the functional equation

$$f(xy) = f(x) + f(y)? \tag{12.6}$$

Since x and y are positive, we can find reals u and v such that $\ln x = u$ and $\ln y = v$, so that $x = e^u$ and $y = e^v$. Then (12.6) becomes $f(e^{u+v}) = f(e^u) + f(e^v)$ so that $f(e^t)$ is a continuous solution to Cauchy's equation. Hence $f(e^t) = kt$, or $f(x) = k \ln x$ for some constant $k \neq 0$. Setting $k = 1/\ln b$ where $b > 0$, $b \neq 1$ yields $f(x) = \ln x/\ln b = \log_b x$. Hence the functional equation (12.6) characterizes logarithmic functions, as they are its only (nontrivial) solutions.

The birth of logarithms

When writing the biography of a function one usually begins with the function, and then discusses its properties. The biography of the logarithm proceeds in the opposite direction, in that it is the functional equation $f(xy) = f(x) + f(y)$ that leads to the functions we know as logarithms.

The need to simplify trigonometric calculations motivated John Napier (1550–1617) to construct two related sequences of numbers, one increasing arithmetically and the other decreasing geometrically. The base of Napier's logarithms was (essentially) $1/e$, and discussions with Henry Briggs (1561–1630) led to logarithms based on powers of 10 in Briggs' *Arithmetica Logarithmica*. Thus was born the inverse of $y = 10^x$.

If we replace sums by arithmetic means in Cauchy's functional equation, we obtain *Jensen's functional equation* (Johan Ludwig William Valdemar Jensen, 1859–1925):

$$f\left(\frac{x+y}{2}\right) = \frac{f(x) + f(y)}{2} \tag{12.7}$$

for $f : \mathbb{R} \to \mathbb{R}$. The solution to (12.7) is a special case of the following theorem where we use a weighted mean rather than the usual arithmetic mean.

Theorem 12.7. *Let $f : \mathbb{R} \to \mathbb{R}$ be a function satisfying one of the three properties in Corollary* 12.4 *and the functional equation*

$$f(\lambda_1 x + \lambda_2 y) = \lambda_1 f(x) + \lambda_2 f(y) \tag{12.8}$$

for some pair of positive numbers λ_1, λ_2 such that $\lambda_1 + \lambda_2 = 1$. Then $f(x) = mx + b$ for some real constants m and b.

Proof. If we set $g(x) = f(x) - f(0)$, then g satisfies (12.8) with $g(0) = 0$. Now replace x and y by x/λ_1 and 0, respectively, in (12.8) written for g rather than f to obtain $g(x) = \lambda_1 g(x/\lambda_1)$. Similarly, $g(y) = \lambda_2 g(y/\lambda_2)$. Thus

$$g(x+y) = g\left(\lambda_1 \frac{x}{\lambda_1} + \lambda_2 \frac{y}{\lambda_2}\right) = \lambda_1 g\left(\frac{x}{\lambda_1}\right) + \lambda_2 g\left(\frac{y}{\lambda_2}\right) = g(x) + g(y).$$

∎

That is, g satisfies Cauchy's equation. Since f satisfies one of the conditions in Corollary 12.4, so does g and $g(x) = mx$ for some constant m. Letting $f(0) = b$, we obtain $f(x) = mx + b$ as the general solution to (12.8).

So far we have considered only functional equations for functions of a single variable. In the proof of the next theorem we solve a functional equation for a function of two variables, based on Problem A1 from the 69th William Lowell Putnam Mathematical Competition in 2008.

Theorem 12.8. *Let $f : \mathbb{R} \times \mathbb{R} \to \mathbb{R}$ be a function satisfying*

$$f(x, y) + f(y, z) + f(z, x) = 0 \tag{12.9}$$

for all real numbers x, y, and z. Then there exists a function $g : \mathbb{R} \to \mathbb{R}$ such that $f(x, y) = g(x) - g(y)$ for all real numbers x and y.

Proof. Setting $x = y = z = 0$ yields $f(0,0) = 0$, setting $y = z = 0$ yields $f(x,0) + f(0,x) = 0$, and setting $z = 0$ yields $f(x,y) + f(y,0) + f(0,x) = 0$. Hence $f(x,y) = f(x,0) - f(0,y)$, so we let $g(x) = f(x,0)$ for all x. The solution is completely general, requiring no assumptions whatsoever about f.

Since the solution to (12.9) implies $f(z,x) = -f(x,z)$, (12.9) is equivalent to $f(x,y) + f(y,z) = f(x,z)$, known as *Sincov's functional equation* (Dmitrii Matveevich Sincov, 1867–1946), which Sincov elegantly solved as follows [Gronau, 2000]: Rewrite the equation as $f(x,y) = f(x,z) - f(y,z)$. The left hand side is independent of z, hence so is the right hand side, and thus we can set $z = a$ and $g(x) = f(x,a)$ to immediately obtain $f(x,y) = g(x) - g(y)$. ∎

Another functional equation for functions of two variables arises when we search for homogeneous functions: A function $f : \mathbb{R} \times \mathbb{R} \to \mathbb{R}$ is *homogeneous of degree k* if for all x, y, z ($z \neq 0$) in $\mathbb{R}$ f satisfies

$$f(xz, yz) = z^k f(x,y), \qquad (12.10)$$

and f is *homogeneous* if it is homogeneous of degree k for some k. Homogeneous functions have the property that if the arguments are scaled by some factor, then the values of the function are scaled by a power of that factor. For example, the arithmetic mean $(x + y)/2$ is homogeneous of degree one, and the polynomial $3x^3 + 5xy^2$ is homogeneous of degree 3. Equation (12.10) is sometimes called *Euler's functional equation for homogeneous functions*, as Euler was the first to find its solutions.

If $x \neq 0$, define $g(u) = f(1,u)$ and observe that

$$f(x,y) = f(x \cdot 1, x \cdot (y/x)) = x^k f(1, (y/x)) = x^k g(y/x).$$

If $x = 0$ and $y \neq 0$, then $f(0,y) = f(y \cdot 0, y \cdot 1) = y^k f(0,1) = y^k c$. If $x = y = 0$, then $f(0,0) = z^k f(0,0)$ with $z \neq 0$, thus $f(0,0) = 0$ for $k \neq 0$ and $f(0,0)$ arbitrary for $k = 0$. It is easy to verify that these functions do indeed satisfy (12.10), and hence we have proved

Theorem 12.9. *The general solution to Euler's functional equation* (12.10) *for homogeneous functions with x, y, z ($z \neq 0$) in $\mathbb{R}$ is*

$$f(x,y) = \begin{cases} x^k g(y/x), & x \neq 0, \\ y^k c, & x = 0, \ y \neq 0, \\ 0, & x = y = 0, \end{cases}$$

for k $\neq$ 0 and

$$f(x, y) = \begin{cases} g(y/x), & x \neq 0, \\ c, & x = 0, \ y \neq 0, \\ \text{arbitrary}, & x = y = 0, \end{cases}$$

for k $=$ 0, where c is an arbitrary constant and g an arbitrary function of a single variable.

12.8   Functional inequalities

Inequalities permeate mathematics, and in this section we examine some functional inequalities and applications related to the functional equations in the preceding section. However, the focus is different: with functional equations we seek solutions, whereas we use functional inequalities to describe properties of certain classes of functions.

We begin with a simple example. If we replace the equals sign in the functional equation $f(\lambda_1 x + \lambda_2 y) = \lambda_1 f(x) + \lambda_2 f(y)$ from Theorem 12.7 with an inequality, we describe convex and concave functions.

Definition 12.1. A function $f : I \to \mathbb{R}$, I an interval of real numbers, is *convex* if f satisfies

$$f(\lambda_1 x + \lambda_2 y) \leq \lambda_1 f(x) + \lambda_2 f(y)$$

for all x in I and all positive λ_1, λ_2 such that $\lambda_1 + \lambda_2 = 1$ and *concave* when the sense of the inequality is reversed.

Geometrically, a function f is convex whenever a chord connecting two points on the graph of the function lies above the graph of the function (and similarly for concave functions). See Figure 12.11. As a consequence of

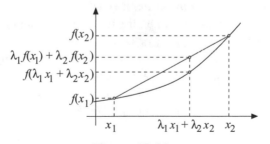

Figure 12.11.

Theorem 12.7, the only functions that are simultaneously convex and concave are the linear functions.

It is easy to extend the inequality in Definition 12.1 to arguments of the form $\lambda_1 x_1 + \lambda_2 x_2 + \cdots + \lambda_n x_n$ for positive $\lambda_1, \lambda_2, \ldots, \lambda_n$ such that $\lambda_1 + \lambda_2 + \cdots + \lambda_n = 1$. The proof requires induction, but it suffices to illustrate the procedure for $n = 3$:

$$
\begin{aligned}
f(\lambda_1 x &+ \lambda_2 y + \lambda_3 z) \\
&= f\left((\lambda_1 + \lambda_2)\left(\frac{\lambda_1 x + \lambda_2 y}{\lambda_1 + \lambda_2}\right) + \lambda_3 z)\right) \\
&\leq (\lambda_1 + \lambda_2) f\left(\frac{\lambda_1 x + \lambda_2 y}{\lambda_1 + \lambda_2}\right) + \lambda_3 f(z) \\
&\leq (\lambda_1 + \lambda_2)\left[\frac{\lambda_1}{\lambda_1 + \lambda_2} f(x) + \frac{\lambda_2}{\lambda_1 + \lambda_2} f(y)\right] + \lambda_3 f(z) \\
&= \lambda_1 f(x) + \lambda_2 f(y) + \lambda_3 f(z).
\end{aligned}
$$

In general we have $f\left(\sum_{i=1}^{n} \lambda_i x_i\right) \leq \sum_{i=1}^{n} \lambda_i f(x_i)$ when f is convex (and with the reversed inequality for f concave). This is *Jensen's inequality*, and is often used in the special case where $\lambda_i = 1/n$ for $i = 1, 2, \ldots, n$. For example, the natural logarithm is concave, hence for positive numbers $x_1, x_2, \ldots, x_n$ and positive numbers $\lambda_1, \lambda_2, \ldots, \lambda_n$ such that $\lambda_1 + \lambda_2 + \cdots + \lambda_n = 1$,

$$
\begin{aligned}
\ln(\lambda_1 x_1 + \lambda_2 x_2 + \cdots + \lambda_n x_n) &\geq \lambda_1 \ln x_1 + \lambda_2 \ln x_2 + \cdots + \lambda_n \ln x_n \\
&= \ln(x_1^{\lambda_1} x_2^{\lambda_2} \cdots x_n^{\lambda_n})
\end{aligned}
$$

which establishes the generalized AM-GM inequality for n numbers.

Another mean, less well known than the arithmetic and geometric, is the *root mean square*, or *quadratic mean*. It is defined as the square root of the arithmetic mean of the squares, i.e., the root mean square of the (positive or negative) numbers $x_1, x_2, \ldots, x_n$ is $[(x_1^2 + x_2^2 + \ldots + x_n^2)/n]^{1/2}$. It is often used to measure magnitude in physics and electrical engineering when quantities are both positive and negative, such as in waveforms. To compare the root mean square to the arithmetic mean, we employ Jensen's inequality to the function $f(x) = x^2$, which is convex on $\mathbb{R}$:

$$
\left(\frac{x_1 + x_2 + \cdots + x_n}{n}\right)^2 \leq \frac{x_1^2 + x_2^2 + \cdots + x_n^2}{n},
$$

and taking square roots and using that for any a in $\mathbb{R}$, $a \leq |a| = \sqrt{a^2}$ we get

$$\frac{x_1 + x_2 + \cdots + x_n}{n} \leq \left| \frac{x_1 + x_2 + \cdots + x_n}{n} \right| \leq \left[\frac{x_1^2 + x_2^2 + \cdots + x_n^2}{n} \right]^{1/2}.$$

Thus the root mean square is at least as large as the arithmetic mean.

Functions that satisfy Cauchy's functional equation (12.4) are often called *additive*, and when the equals sign in (12.4) is replaced by an inequality, we describe *subadditive* and *superadditive* functions:

Definition 12.2. Let A and B be subsets of $\mathbb{R}$ closed under addition. A function $f : A \to B$ is *subadditive* if for all x and y in A we have

$$f(x + y) \leq f(x) + f(y), \tag{12.11}$$

and *superadditive* when the inequality is reversed.

For example, $|x|$ and e^x are subadditive on $\mathbb{R}$, and $\sqrt{x}$ is subadditive on $[0, \infty)$. Since f is superadditive if and only if $-f$ is subadditive, we restrict our attention to subadditive functions.

We now present some properties of subadditive functions with domain $\mathbb{R}$ or $[0, \infty)$.

Theorem 12.10. *Let $f : \mathbb{R} \to \mathbb{R}$ be a subadditive function and n a positive integer. Then we have*

(a) $f(0) \geq 0$,

(b) $f(nx) \leq nf(x)$,

(c) $f(x/n) \geq f(x)/n$,

(d) $f(-x) \geq -f(x)$,

(e) *if f is even, then $f(x) \geq 0$, and*

(f) *if f is odd, then f is additive.*

Proof. Part (a) follows from (12.11) with $x = y = 0$. Part (b) holds for $n = 1$ and by induction for $n > 1$ since

$$f((n + 1)x) \leq f(nx) + f(x) \leq nf(x) + f(x) = (n + 1)f(x).$$

Part (c) follows from (b) by replacing x by x/n. Part (d) follows from (a) and (12.11):

$$0 \le f(0) = f(x - x) \le f(x) + f(-x).$$

When f is even $f(-x) = f(x)$, so part (e) follows from (d): $f(x) = f(-x) \ge -f(x)$ so $2f(x) \ge 0$. When f is odd $f(-y) = -f(y)$ so part (f) follows from (12.11):

$$f(x) = f(x + y - y) \le f(x + y) + f(-y) = f(x + y) - f(y),$$

which when combined with (12.11) yields the additivity of f. ∎

Theorem 12.11. *If $f : [0, \infty) \to [0, \infty)$ is concave, then it is subadditive. If f is convex, then it is superadditive.*

Proof. First observe that $f(0) \ge 0$. Then for all a and b such that $a + b \ne 0$,

$$f(a) = f\left(\frac{a}{a+b}(a + b) + \frac{b}{a+b}0\right) \ge \frac{a}{a+b}f(a + b) + \frac{b}{a+b}f(0),$$

$$f(b) = f\left(\frac{a}{a+b}0 + \frac{b}{a+b}(a + b)\right) \ge \frac{a}{a+b}f(0) + \frac{b}{a+b}f(a + b).$$

Adding the inequalities (and recalling that $f(0) \ge 0$) yields

$$f(a) + f(b) \ge f(a + b) + f(0) \ge f(a + b).$$

If $a + b = 0$, then $a = b = 0$ and thus

$$f(a) + f(b) = 2f(0) \ge f(0) = f(a + b). \qquad ∎$$

Similarly one shows that convexity implies superadditivity.

12.9 Euler's series for $\pi^2/6$

> *Even as the finite encloses an infinite series*
> *And in the unlimited limits appear,*
> *So the soul of immensity dwells in minutia*
> *And in the narrowest limits no limits inhere.*
> *What joy to discern the minute in infinity!*
>
> Jakob Bernoulli, *Ars Conjectendi*

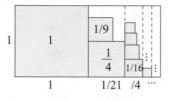

Figure 12.12.

One of the most celebrated results of Leonhard Euler (1707–1783) is

Theorem 12.12. *The series of reciprocals of squares converges to $\pi^2/6$, i.e.,*

$$1 + \frac{1}{2^2} + \frac{1}{3^2} + \frac{1}{4^2} + \cdots = \frac{\pi^2}{6}. \tag{12.12}$$

Before presenting an elegant proof of this result using only multivariable calculus [Harper, 2003], we note that it is easy to see that the series converges to a value less than 2. In Figure 12.12 we see that squares with areas 1, 1/4, 1/9, and so on fit easily into a 1-by-2 rectangle, and hence the increasing sequence of partial sums of the series (12.12) is bounded above.

It is tempting to try to fit the squares with areas 1, 1/4, 1/9, ... into, say, a $\pi/2$-by-$\pi/3$ rectangle (by cutting some of the squares into smaller pieces), but no one has done that yet. So we present an analytic proof.

Proof. To establish (12.12), we need only show that

$$\sum_{n=0}^{\infty} \frac{1}{(2n+1)^2} = \frac{\pi^2}{8}, \tag{12.13}$$

since

$$\frac{3}{4} \sum_{n=1}^{\infty} \frac{1}{n^2} = \sum_{n=1}^{\infty} \frac{1}{n^2} - \sum_{n=1}^{\infty} \frac{1}{(2n)^2} = \sum_{n=0}^{\infty} \frac{1}{(2n+1)^2}.$$

Consider the double integral

$$\int_0^{\infty} \int_0^1 \frac{x}{(x^2+1)(x^2y^2+1)} \, dy \, dx = \int_0^{\infty} \left[\frac{\arctan xy}{x^2+1} \right]_0^1 dx$$

$$= \int_0^{\infty} \frac{\arctan x}{x^2+1} \, dx = \left. \frac{(\arctan x)^2}{2} \right|_0^{\infty}$$

$$- \frac{\pi^2}{8}.$$

Reversing the order of integration yields

$$\int_0^1 \int_0^\infty \frac{x}{(x^2+1)(x^2y^2+1)} \, dx\,dy$$

$$= \int_0^1 \int_0^\infty \frac{1}{2(y^2-1)} \left[\frac{2xy^2}{x^2y^2+1} - \frac{2x}{x^2+1} \right] dx\,dy,$$

$$= \int_0^1 \frac{1}{2(y^2-1)} \left[\ln\left(\frac{x^2y^2+1}{x^2+1} \right) \right]_0^\infty dy,$$

$$= \int_0^1 \frac{\ln y^2}{2(y^2-1)} \, dy = \int_0^1 \frac{\ln y}{y^2-1} \, dy.$$

We now integrate by parts using $u = \ln y$ and $dv = dy/(y^2-1)$:

$$\int_0^1 \frac{\ln y}{y^2-1} \, dy = \frac{1}{2} \ln y \ln \frac{1-y}{1+y} \Big|_0^1 - \frac{1}{2} \int_0^1 \frac{1}{y} \ln \frac{1-y}{1+y} \, dy$$

$$= \frac{1}{2} \int_0^1 \frac{1}{y} \ln \frac{1+y}{1-y} \, dy.$$

Finally we expand the integrand into its Maclaurin series and switch the order of integration and summation:

$$\frac{1}{2} \int_0^1 \frac{1}{y} \ln \frac{1+y}{1-y} \, dy = \int_0^1 \frac{1}{y} \sum_{n=0}^\infty \frac{y^{2n+1}}{2n+1} \, dy$$

$$= \sum_{n=0}^\infty \int_0^1 \frac{y^{2n}}{2n+1} \, dy = \sum_{n=0}^\infty \frac{1}{(2n+1)^2},$$

which proves (12.13). ∎

There are many other proofs of (12.12). For collections of them see [Kalman, 1993] and [Chapman, 2003].

Two positive integers are *relatively prime* if their greatest common divisor is 1. A surprising result is

Corollary 12.4. *The probability that two positive integers are relatively prime is $6/\pi^2$.*

Let (a, b) denote the greatest common divisor of a and b, and let p be the probability that $(a, b) = 1$. There are a number of proofs that $p = 6/\pi^2$, most require a substantial amount of number theory. The following simple proof is from [Abrams and Paris, 1992].

Proof. We first show that the probability that $(a, b) = k$ for $k = 1, 2, 3, \ldots$ is p/k^2. Since the probability that k divides a is $1/k$, the probability that k divides both a and b is $1/k^2$. The probability that no multiple of k divides a and b is equal to the probability that $(a/k, b/k) = 1$, which is p. Hence the probability that $(a, b) = k$ is p/k^2.

Since any pair of positive integers must have a greatest common divisor, the sum of the probabilities that $(a, b) = k$ for $k = 1, 2, 3, \ldots$ must be 1, so that

$$1 = \sum_{k=1}^{\infty} \frac{p}{k^2} = p \sum_{k=1}^{\infty} \frac{1}{k^2} = p \frac{\pi^2}{6}.$$

Thus $p = 6/\pi^2 \approx .608$. ∎

We must be careful when speaking of choosing numbers at random from infinite sets such as $\{1, 2, 3, \ldots\}$. The probability p that $(a, b) = 1$ is usually defined as the limit as $n \to \infty$ of the probability p_n that $(a, b) = 1$ when a and b are restricted to the range $1 \le a \le n$ and $1 \le b \le n$. See [Yaglom and Yaglom, 1964] for details.

12.10 The Wallis product

In 1656 John Wallis (1616–1703) published the following curious but remarkable formula expressing $\pi/2$ as an infinite product:

Theorem 12.13.

$$\frac{\pi}{2} = \frac{2}{1} \cdot \frac{2}{3} \cdot \frac{4}{3} \cdot \frac{4}{5} \cdot \frac{6}{5} \cdot \frac{6}{7} \cdot \frac{8}{7} \cdot \frac{8}{9} \cdots .$$

Proof. To prove this, we will use integral calculus [Taylor and Mann, 1972], a tool Wallis did not have at his disposal. We will show that

$$\lim_{n \to \infty} \left[\frac{2}{1} \cdot \frac{4}{3} \cdot \cdots \cdot \frac{2n}{2n-1} \right]^2 \frac{1}{2n+1} = \frac{\pi}{2}. \qquad (12.14)$$

Let $I_n = \int_0^{\pi/2} \sin^n x \, dx$ for a nonnegative integer n. Integration by parts yields $I_n = ((n+1)/n) I_{n-2}$, which, along with $I_0 = \pi/2$ and $I_1 = 1$, yields

$$I_{2n} = \frac{2n-1}{2n} \cdot \frac{2n-3}{2n-2} \cdot \cdots \cdot \frac{3}{4} \cdot \frac{1}{2} \cdot \frac{\pi}{2}$$

and

$$I_{2n+1} = \frac{2n}{2n+1} \cdot \frac{2n-2}{2n-1} \cdot \cdots \cdot \frac{2}{3} \cdot 1.$$

But for $0 < x < \pi/2$, $\sin^{n+1} x < \sin^n x$ and hence $I_{n+1} < I_n$. So $I_{2n+1} < I_{2n}$ implies

$$\frac{2}{1} \cdot \frac{4}{3} \cdot \ \cdots \ \cdot \frac{2n}{2n-1} \cdot \frac{1}{2n+1} < \frac{1}{2} \cdot \frac{3}{4} \cdot \ \cdots \ \cdot \frac{2n-1}{2n} \cdot \frac{\pi}{2},$$

or

$$\left[\frac{2}{1} \cdot \frac{4}{3} \cdot \ \cdots \ \cdot \frac{2n}{2n-1} \right]^2 \frac{1}{2n+1} < \frac{\pi}{2}.$$

Similarly $I_{2n} < I_{2n-1}$ implies

$$\frac{2n}{2n+1} \cdot \frac{\pi}{2} < \left[\frac{2}{1} \cdot \frac{4}{3} \cdot \ \cdots \ \cdot \frac{2n}{2n-1} \right]^2 \frac{1}{2n+1},$$

and so we have

$$\frac{2n}{2n+1} \cdot \frac{\pi}{2} < \left[\frac{2}{1} \cdot \frac{4}{3} \cdot \ \cdots \ \cdot \frac{2n}{2n-1} \right]^2 \frac{1}{2n+1} < \frac{\pi}{2}.$$

The result follows by using the squeeze theorem for limits. ∎

As a practical tool for calculating π, Wallis's formula is useless, due to the extremely slow convergence of the infinite product. Computing the product in (12.14) for $n = 500$ gives only $\pi \approx 3.13989$. However, the formula is useful in developing approximations. We will use the Wallis product in our derivation of Stirling's approximation to $n!$ in the next section.

12.11   Stirling's approximation of $n!$

Stirling's formula (James Stirling, 1692–1779) is an approximation to $n!$, usually written as

Theorem 12.14 (Stirling's formula). $n! \sim \sqrt{2\pi n} \cdot n^n e^{-n}.$

For two sequences $\{a_n\}$ and $\{b_n\}$, $a_n \sim b_n$ means $\lim_{n \to \infty} a_n/b_n = 1$. We will give a calculus-based proof by showing first that $n! \sim C \sqrt{n} \ n^n e^{-n}$ for some constant C and then use Wallis's product to show that $C = \sqrt{2\pi}$. Our proof is adapted from the one in [Coleman, 1951].

Proof. We begin by using the trapezoidal rule to approximate $\int_1^n \ln x \ dx$, using $n - 1$ trapezoids each with a base of width 1. The trapezoidal rule approximation is

$$\frac{1}{2}[\ln 1 + \ln 2] + \frac{1}{2}[\ln 2 + \ln 3] + \cdots + \frac{1}{2}[\ln(n-1) + \ln n] = \ln n! - \frac{1}{2}\ln n,$$

and the exact value of the integral is $x \ln x - x|_1^n = n \ln n - n + 1$. Hence

$$s_n = \int_1^n \ln x \, dx - \ln n! + \frac{1}{2} \ln n = \left(n + \frac{1}{2}\right) \ln n - n + 1 - \ln n!$$

represents the area under the graph of $y = \ln x$ over $[1,n]$ minus the trapezoidal rule approximation to that area. Since graph of $y = \ln x$ is concave,
the trapezoids lie below the graph, so each s_n is positive, and clearly the
sequence $\{s_n\}$ is increasing in n.

To show that the increasing sequence $\{s_n\}$ has a limit we need only show
that it is bounded above, which we do with a modified midpoint rule approximation to the integral

$$\int_1^n \ln x \, dx = \int_1^{3/2} \ln x \, dx + \int_{3/2}^{n-1/2} \ln x \, dx + \int_{n-1/2}^n \ln x \, dx.$$

In the interval $[1, 3/2]$ we have $\ln x < x - 1$, in the interval $[3/2, n - 1/2]$ we
use the midpoint rule to approximate the integral (the midpoint rule overesti-
mates the integral since the graph of $y = \ln x$ is concave), and in the interval
$[n - 1/2, n]$, $\ln x < \ln n$. Thus we have

$$\int_1^{3/2} \ln x \, dx < \int_1^{3/2} (x - 1) \, dx = 1/8,$$

$$\int_{3/2}^{n-1/2} \ln x \, dx < \ln 2 + \ln 3 + \cdots + \ln(n - 1) = \ln n! - \ln n,$$

$$\int_{n-1/2}^n \ln x \, dx < \int_{n-1/2}^n \ln n \, dx = \frac{1}{2} \ln n,$$

and hence

$$\int_1^n \ln x \, dx = n \ln n - n + 1 < \frac{1}{8} + \ln n! - \frac{1}{2} \ln n,$$

or $s_n < 1/8$. Hence the sequence $\{s_n\}$ has a limit c, which satisfies $0 < c \leq 1/8$.

Now set

$$S_n = \exp(1 - s_n) = \frac{n!}{\sqrt{n} \, n^n e^{-n}}.$$

The sequence $\{S_n\}$ has a limit $C = e^{1-c}$, or $n! \sim C \sqrt{n} \, n^n e^{-n}$. Note that
C lies in the interval $[e^{7/8}, e] \approx [2.40, 2.72]$, i.e., C may be near 2.5.

To show that $C = \sqrt{2\pi}$, we use the following clever argument [Taylor and Mann, 1972]. Consider the ratio

$$\frac{S_n^2}{S_{2n}} = \frac{(n!)^2 e^{2n}}{n^{2n+1}} \cdot \frac{(2n)^{2n+1/2}}{(2n)! e^{2n}}$$

$$= \frac{(n!)^2 2^{2n} \sqrt{2}}{(2n)! \sqrt{n}}$$

$$= \frac{2 \cdot 4 \cdot 6 \cdots (2n)}{1 \cdot 3 \cdot 5 \cdots (2n-1)} \sqrt{\frac{2}{n}}$$

$$= \frac{2 \cdot 4 \cdot 6 \cdots (2n)}{1 \cdot 3 \cdot 5 \cdots (2n-1)} \cdot \frac{1}{\sqrt{2n+1}} \cdot \sqrt{\frac{2(2n+1)}{n}}.$$

Using (12.14), the limit on the right as $n \to \infty$ is $2\sqrt{\pi/2} = \sqrt{2\pi}$, while the limit of S_n^2/S_{2n} is $C^2/C = C$. Hence $C = \sqrt{2\pi}$, which completes the proof. ∎

12.12   Challenges

12.1 Using the fact that $\mathbb{Q}_+$ is denumerable, show that $\mathbb{Q}$ is denumerable.

12.2 Prove that there exists a denumerable set S such that S has nondenumerably many subsets, the intersection of any two of which is finite.

12.3 Use (a) the AM-GM inequality and (b) the Cauchy-Schwarz inequality to give one-line proofs that for positive numbers $a_1, a_2, \ldots, a_n$,

$$(a_1 + a_2 + \cdots + a_n) \left(\frac{1}{a_1} + \frac{1}{a_2} + \cdots + \frac{1}{a_n} \right) \geq n^2.$$

12.4 Prove that the Cauchy-Schwarz inequality for n numbers implies the AM-GM inequality for 2 numbers.

12.5 Show that the AM-GM inequality for n numbers implies $e^{1/e} \geq x^{1/x}$ for $x > 0$.

12.6 Show that for functions $f : \mathbb{R} \to \mathbb{R}$ such that $f(0) = 0$, the equation

$$f(x + y) + f(x - y) = 2f(x)$$

is equivalent to Cauchy's functional equation (12.4).

12.7 What continuous functions $f : (0, \infty) \to \mathbb{R}$ (other than $f(x) \equiv 0$) are completely multiplicative, that is, satisfy the functional equation $f(xy) = f(x)f(y)$?

12.8 Let a, b, c denote the sides of triangle ABC, and R the radius of its circumscribed circle. Prove that $a + b + c \leq 3\sqrt{3}R$. (Hint: the sine function is concave on $[0, \pi]$.)

12.9 Let $f : \mathbb{R} \to \mathbb{R}$ be a subadditive function. (a) Show that if $f(x) \leq x$ for all x, then $f(x) = x$. (b) Does the same conclusion hold if $f(x) \geq x$ for all x?

12.10 The harmonic series (Lemma 1.2) and Euler's series for $\pi^2/6$ (Theorem 12.12) are special cases of the *Riemann zeta function*, $\zeta(s) = \sum_{n=1}^{\infty} n^{-s}$, i.e., $s = 1$ and $s = 2$. Values of $\zeta(s)$ are known when s is an even positive integer, but no simple expression is known for s an odd positive integer larger than 1. Nonetheless, prove that

$$\sum_{k=2}^{\infty} [\zeta(k) - 1] = 1.$$

12.11 Show that

$$\binom{2n}{n} \frac{1}{2^{2n}} \sim \frac{1}{\sqrt{\pi n}}.$$

As a consequence, the probability of obtaining exactly n heads and n tails in $2n$ tosses of a fair coin is approximately $1/\sqrt{\pi n}$.

12.12 Another formula for $n!$ is *Burnside's formula:*

$$n! \sim \sqrt{2\pi}(n + 1/2)^{n+1/2}e^{-(n+1/2)}.$$

Burnside's formula is about twice as accurate as Stirling's. Prove it.

12.13 Show that $1 \cdot 3 \cdot 5 \cdots (2n - 1) \sim \sqrt{2}(2n/e)^n$.

Solutions to the Challenges

Many of the Challenges in this book have multiple solutions. Here we give but one simple solution to each Challenge, and encourage readers to search for others that may be more charming.

Chapter 1

1.1 (a) See Figure S1.1a, and count the objects on the rising (or falling) diagonals.

 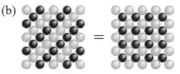

Figure S1.1.

(b) See Figure S1.1b.

(c) See Figure S1.2 [Arcavi and Flores, 2000].

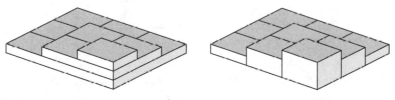

Figure S1.2.

1.2 See Figure S1.3.

1.3 For the basis step, $1 + 9 = 10 = t_{1+3}$. Assume $1 + 9 + 9^2 + \cdots + 9^n = t_{1+3+3^2+\cdots+3^n}$. Then $1 + 9 + 9^2 + \cdots + 9^{n+1} = 9(1 + 9 + 9^2 + \cdots + 9^n) + 1 = 9(t_{1+3+3^2+\cdots+3^n}) + 1 - t_{3(1+3+3^2+\cdots+3^n)+1} = t_{1+3+3^2+\cdots+3^{n+1}}$.

239

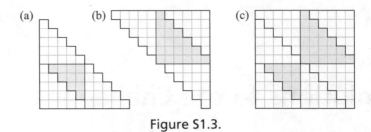

Figure S1.3.

1.4 Since $t_{n-1} + n = t_n$, replacing n by t_n yields $t_{t_n-1} + t_n = t_{t_n}$.

1.5 No. Since $N_k = p_1 p_2 \cdots p_k$ has the form $2(2k + 1)$, Euclid numbers have the form $4k + 3$. However, all odd squares are of the form $4k + 1$.

1.6 See Figure S1.4 [Ollerton, 2008].

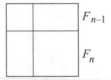

Figure S1.4.

1.7 See Figure S1.5.

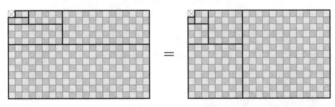

Figure S1.5.

1.8 See Figure S1.6 to establish $F_{n+1}^2 - F_n F_{n+2} = F_{n-1} F_{n+1} - F_n^2$.

Thus the terms of the sequence $\{F_{n-1} F_{n+1} - F_n^2\}_{n=2}^{\infty}$ all have the same magnitude and simply alternate in sign. So we need only evaluate the base case $n = 2$: $F_1 F_3 - F_2^2 = 1$, so $F_{n-1} F_{n+1} - F_n^2$ is $+1$ when n is even and -1 when n is odd, i.e., $F_{n-1} F_{n+1} - F_n^2 = (-1)^n$.

1.9 In base 10, $\{a_n\} = \{1, 2, 5, 10, 21, 42, \cdots\}$, but in base 4, $\{a_n\} = \{1, 2, 11, 22, 111, 222, \cdots\}$. So it appears that

$$a_{2n-1} = 4^{n-1} + 4^{n-2} + \cdots + 4 + 1 = \frac{4^n - 1}{3} \quad \text{and} \quad a_{2n} = 2\frac{4^n - 1}{3}.$$

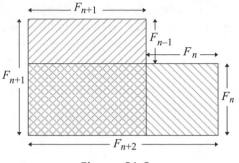

Figure S1.6.

This is readily verified by substitution into the given recurrence.

1.10 The theorem clearly holds for $n = 0$, For $p = 2$, note that $n^2 - n = n(n-1)$ is always even. For p odd, note that $(-n)^p - (-n) = -(n^p - n)$.

1.11 Each factorization of n as a product ab with $a \neq b$ yields two distinct factors, one less than $\sqrt{n}$ and one more than $\sqrt{n}$. So if n is not a square, then $\tau(n)$ is even. If n is a square, then there is a factorization ab with $a = b$, in which case we obtain one more factor.

1.12 Since n is an even perfect number greater than 6, $n = 2^{p-1}(2^p - 1) = t_{2^p - 1}$ where p is an odd prime. But $2^p \equiv 2 \pmod 3$ so that $2^p = 3k + 2$ for some integer k, and hence $2^p - 1 = 3k + 1$. So from Lemma 1.1b we have $n = t_{2^p-1} = t_{3k+1} = 9t_k + 1$, which establishes (a). For (b), write $(n - 1)/9 = (k + 1) \cdot (k/2)$ and observe that $3k = 2^p - 2 = 2(2^{p-1} - 1)$, so that k must be even and hence $k + 1$ is odd. Furthermore $k/2$ is an integer, and $3 \cdot (k/2) = 2^{p-1} - 1$, so $k/2$ must be an odd integer. Thus $(n - 1)/9$ is a product of two odd numbers $k + 1$ and $k/2$, and when $n > 28$, both $k + 1$ and $k/2$ are greater than 1.

Chapter 2

2.1 (a) Set $\sqrt{3} = m/n$ and construct an integer-sided right triangle with sides of length n, m, and $2n$. Using the same construction as in Figure 1.1, the smaller triangle has sides n, $m/3$, and $2m/3$. But $m/3$ is an integer, since $m^2 = 3n^2$ implies that m is divisible by 3.

(b) Set $\sqrt{5} = m/n$ and construct an integer-sided right triangle with sides of length n, $2n$, and m. Using the same construction as in

Figure 1.1, the smaller triangle has sides $m - n$, $(m - n)/2$, and $2n - (m - n)/2$. But $(m - n)/2$ is an integer since both m and n must be odd (m and n cannot both be even, and since $m^2 = 3n^2$ it is impossible for one of m and n to be even and the other odd).

2.2 Use the fact that squares are congruent to 0 or 1 mod 3.

2.3 Assume $\sqrt[n]{2}$ is rational, i.e., there exist positive integers p and q such then $\sqrt[n]{2} = p/q$. Then $q^n + q^n = p^n$, contradicting Fermat's Last Theorem [Schultz, 2003].

2.4 Let x denote the length of the diagonal, as illustrated in Figure S2.1a.

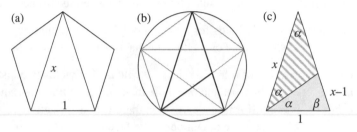

Figure S2.1.

Two diagonals from a common vertex form an isosceles triangle, which can be partitioned by a segment of another diagonal into two smaller triangles, as illustrated in Figure S2.1b and S2.1c. The angles labeled α in Figure S2.1c are equal since each subtends an arc equal to 1/5 the circumference of the circumscribing circle. Thus the striped triangle is also isosceles, $\beta = 2\alpha$, and so $\alpha = 36°$, since the sum of the angles in the original triangle yields $5\alpha = 180°$. It now follows that the unlabeled angle in the gray triangle is $\beta = 72°$, and hence the triangle is also isosceles and similar to the original triangle. Hence $x/1 = 1/(x - 1)$, and so x is the (positive) root of $x^2 - x - 1 = 0 = 0$. That is, $x = \varphi$ as claimed.

2.5 Since the angles of the two trapezoids are the same (60° and 120°), we need only find the value of x that makes the lengths of the sides proportional. The lengths of the sides are indicated in Figure S2.2 (recall that the triangles are equilateral).

Then $\dfrac{2x - 1}{1 - x} = \dfrac{x}{1} = \dfrac{1 - x}{x}$ if and only if $x^2 + x - 1 = 0$, i.e., $x = 1/\varphi$.

Figure S2.2.

2.6 Since $A = \pi(a^2 - b^2)$ and $E = \pi a b$, $A = E$ if and only if $a^2 - ab - b^2 = 0$, or equivalently, $(a/b)^2 - (a/b) - 1 = 0$. Since $a/b > 0$, a/b must be the golden ratio. Such an ellipse is sometimes called a *golden ellipse*, as it can be inscribed in a golden rectangle.

2.7 Introduce an xyz coordinate system on the rectangles so that the origin is at the center and each rectangle lies in one of the coordinate planes. The coordinates of the twelve vertices are $(0, \pm a/2, \pm b/2)$, $(\pm b/2, 0, \pm a/2)$, and $(\pm a/2, \pm b/2, 0)$. The faces are equilateral triangles if and only if

$$\sqrt{\left(\frac{b}{2} - \frac{a}{2}\right)^2 + \left(0 + \frac{b}{2}\right)^2 + \left(\frac{a}{2} - 0\right)^2} = a,$$

i.e., $(b/a)^2 = 1 + (b/a)$ and thus $b/a = \varphi$.

2.8 From Binet's formula we have $F_n = (\varphi^n/\sqrt{5}) - (-1)^n(1/\varphi^n\sqrt{5})$. Since $\varphi^n\sqrt{5} \geq \varphi\sqrt{5} > 3$, F_n (an integer) is within $1/3$ of $\varphi^n/\sqrt{5}$, hence it is the nearest integer.

2.9 Yes. If the sequence is geometric, then for some $r \neq 0$, the terms have the form ar^{n-1} and for $n \geq 3$, $ar^n = ar^{n-1} + ar^{n-2}$. This implies that $r^2 = r + 1$, and thus r equals φ or $-1/\varphi$.

2.10 Evaluating the integral suffices since the integrand is positive on $[0,1]$, and thus $22/7 - \pi > 0$. So we have

$$\int_0^1 \frac{x^4(1-x)^4}{1+x^2}\,dx = \int_0^1 \left(x^6 - 4x^5 + 5x^4 - 4x^2 + 4 - \frac{4}{1+x^2}\right)dx$$

$$= \frac{x^7}{7} - \frac{2x^6}{3} + x^5 - \frac{4x^3}{3} + 4x - 4\arctan x\Big|_0^1$$

$$= \frac{1}{7} - \frac{2}{3} + 1 - \frac{4}{3} + 4 - \pi = \frac{22}{7} - \pi.$$

2.11 (a) In Figure 2.9, compare the slope of the secant line to the slope of
            the tangent line at its left endpoint.

      (b) Clearing fractions in the inequality from (a) yields

$$a^n[(n+1)b - na] < b^{n+1}. \tag{S2.1}$$

Setting $a = 1 + (1/(n+1))$ and $b = 1 + (1/(n))$ yields

$$\left(1 + \frac{1}{n+1}\right)^n \left[\left(\frac{n+2}{n+1}\right)^2 \frac{n(n+2)^2 + 1}{n(n+2)^2}\right] < \left(1 + \frac{1}{n}\right)^{n+1}.$$

But $(1 + 1/(n+1))^{n+2}$ is less than the number on the left side of
the above inequality, thus $\{(1 + 1/n)^{n+1}\}$ is decreasing. Setting
$a = 1$ and $b = 1 + (1/n)$ in (S2.1) yields

$$2 < \frac{2n+1}{n} < \left(1 + \frac{1}{n}\right)^{n+1}$$

and hence $\{(1 + 1/n)^{n+1}\}$ is bounded below.

      (c) The difference between $(1+1/n)^{n+1}$ and $(1+1/n)^n$ is $(1 + 1/n)^n/$
            n, which goes to 0 as $n \to \infty$.

2.12 From Challenge 2.9(c) we have

$$\prod_{k=1}^{n} \frac{(k+1)^k}{k^k} < e^n < \prod_{k=1}^{n} \frac{(k+1)^{k+1}}{k^{k+1}},$$

which simplifies to

$$\frac{(n+1)^n}{n!} < e^n < \frac{(n+1)^{n+1}}{n!}.$$

Hence

$$\frac{n+1}{e} < \sqrt[n]{n!} < \frac{(n+1)^{1+1/n}}{e},$$

and thus

$$\left(\frac{n+1}{n}\right) \cdot \frac{1}{e} < \frac{\sqrt[n]{n!}}{n} < \left(\frac{n+1}{n}\right) \cdot \frac{(n+1)^{1/n}}{e}.$$

But $(n+1)^{1/n} \to 1$ and $(n+1)/n \to 1$ as $n \to \infty$, thus $\lim_{n \to \infty}$
$\sqrt[n]{n!}/n = 1/e$ [Schaumberger, 1984].

2.13 From Section 2.10 we have $e^{1/e} > \pi^{1/\pi}$, hence $e^{\pi} > \pi^{e}$.

2.14 (a) If $2^{\sqrt{2}}$ is irrational, then it is our example. If $2^{\sqrt{2}}$ is rational, then $(2^{\sqrt{2}})^{1/(2\sqrt{2})} = \sqrt{2}$ is our example.

(b) $1^{\sqrt{2}} = 1$ is an example. As a consequence of the solution to Hilbert's seventh problem, we cannot ask for an example showing that a rational number not 0 or 1 to an irrational power may be rational.

(c) Because it is too easy: $\sqrt{2}^{1} = \sqrt{2}$ and $\sqrt{2}^{2} = 2$.

2.15 The values of $\gamma - a_n$ can be represented by a picture similar to Figure 2.11, but over the interval $[n+1, \infty)$. The convexity of the graph of $y = 1/x$ then insures that the sum of the shaded regions is at least half the area of the rectangle with base 1 and height $1/(n+1)$, thus $\gamma - a_n > 1/2(n+1)$.

2.16 We use induction to show that $\{x_n\}$ converges. The sequence $\{x_n\}$ is increasing since $x_2 > x_1$ and $x_n > x_{n-1}$ implies

$$x_{n+1} = \sqrt{k + x_n} > \sqrt{k + x_{n-1}} = x_n$$

and $\{x_n\}$ is bounded since $x_1 < \sqrt{k} + 1$ and $x_n < \sqrt{k} + 1$ implies

$$x_{n+1} = \sqrt{k + x_n} < \sqrt{k + \sqrt{k} + 1} < \sqrt{k + 2\sqrt{k} + 1} = \sqrt{k} + 1.$$

So if $\lim_{n \to \infty}\{x_n\} = x$, then $x^2 = k + x$ and the positive root of this equation is $x = (1 + \sqrt{1 + 4k})/2$.

Chapter 3

3.1 No. If an equilateral lattice triangle S has side length s, then s^2 is an integer by the Pythagorean theorem. Hence the area $A(S) = s^2\sqrt{3}/4$ is irrational, however the area of ever lattice polygon is rational by Pick's theorem.

3.2 No. The volume of the tetrahedron in the hint is $k/6$, yet it has only four lattice points (the vertices) on its surface or in its interior.

3.3 Consider the set of points in the Cartesian plane with integer coordinates (the *lattice* points of the plane). Any line passing through two lattice points will pass through infinitely many lattice points.

3.4 Choose the seven points to be the vertices, the midpoints of the sides, and the center of an equilateral triangle.

3.5 The vertical sides of a rectangle can be chosen in $\binom{m}{2}$ ways and the horizontal sides in $\binom{n}{2}$ ways, hence there are $\binom{m}{2}\binom{n}{2}$ such rectangles.

3.6 Assume $n = 4k - 1$ and $r_2(n) > 0$. If $x^2 + y^2 = n$ for an odd n, then one of x and y is even while the other is odd. Hence one of x^2 and y^2 is a multiple of 4 while the other is one more than a multiple of 4. Thus an odd n must have the form $n = 4k + 1$ rather than $n = 4k - 1$.

3.7 Proceed as in the proof of Theorem 3.2 but with spheres and cubes in space to show that

$$\frac{4}{3}\pi(\sqrt{n} - \sqrt{3}/2)^3 < N_3(n) < \frac{4}{3}\pi(\sqrt{n} + \sqrt{3}/2)^3.$$

Dividing by $n\sqrt{n}$ and taking limits yields $\lim_{n\to\infty} N_3(n)/n\sqrt{n} = 4\pi/3$.

3.8 Since there are five vertices and two colors, it follows from the pigeon-hole principle that there are at least three vertices A, B, and C in the same color, say white. Two cases occur: Case I: Vertices A, B and C are adjacent (Figure S3.1a). Then $BA = BC$. Case II: Two of the vertices (say A and B) are adjacent, and the third is opposite to them (Figure S3.1b). Then $CA = CB$ [Bankov, 1995].

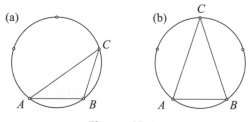

Figure S3.1.

3.9 Write each of the selected integers in the form $2^k q$, where q is an odd integer. Place two integers in the same pigeonhole if each has the same value for q. Since there are only n odd integers between 1 and $2n$, at least two of the integers will be in the same pigeonhole, and the one with the smaller value of k will divide the other.

3.10 (a) Consider an equilateral triangle with side length one inch. At least two of the three vertices must have been painted the same color.

(b) Assume not, i.e., assume every pair of points one inch apart have different colors. Consider two equilateral triangles ABC and $A'BC$ with edges one inch long sharing a common edge BC. Then the vertices of each must all be different colors, so A and A', $\sqrt{3}$ inches apart, have the color. This is true for any pair of points $\sqrt{3}$ inches apart. Now consider an isosceles triangle DEF with $DE = DF = \sqrt{3}$ and $EF = 1$ inch. D and E, and D and F must be the same color, so E and F are the same color, a contradiction.

(c) No. Tile the plane with congruent 3-by-3 arrays of squares 0.6 inches on a side, painted with the nine different colors. Two points within a square are less than one inch apart, whereas two points in different squares of the same color are more than one inch apart. [Parts (b) and (c) of this problem are from Problem A4 on the 1988 William Lowell Putnam Competition.]

3.11 Consider seven collinear points. Some four of them P_1, P_2, P_3, P_4 must be the same color, say red. Project them onto two lines parallel to the first line, yielding two quadruples $(Q_1, Q_2, Q_3, Q_4), (R_1, R_2, R_3, R_4)$ of points. See Figure S3.2.

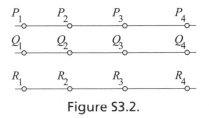

Figure S3.2.

If any two of the Qs are red, an all red rectangle $P_i P_j Q_j Q_i$ exists, similarly is any two Rs are red. If neither of these cases occur, then three or more of the Qs and three or more of the Rs are blue, in which case there must be an all-blue rectangle with Q- and R-vertices [Honsberger, 1978].

Chapter 4

4.1 Let r be the radius of the circle. Then the circumradius of the inscribed n-gon is r, the circumradius of the circumscribed n-gon is $r/(\cos \pi/n)$, and the result follows.

4.2 Let s denote the side length of the inscribed n-gon. Then $p_n = ns$ and $a_{2n} = 2n \cdot (1/2) \cdot 1 \cdot (s/2) = ns/2$, so $a_{2n} = p_n/2$.

4.3 Let R denote the radius of the circle, let s_n and S_n denote the side lengths of the regular inscribed and circumscribed n-gons, respectively, and let r_n denote the inradius of an inscribed n-gon (see Figure S4.1a). Then $S_n/s_n = R/r_n$. Similar triangles within the inscribed n- and $2n$-gons (see Figure S4.1b) yields $(s_{2n}/2)/r_{2n} = (s_n/2)/(R + r_n)$ or $s_{2n}/r_{2n} = s_n/(R + r_n)$. Thus

$$\frac{2p_n P_n}{p_n + P_n} = \frac{2ns_n S_n}{s_n + S_n} = \frac{2ns_n^2 R/r_n}{s_n + s_n R/r_n}$$

$$= 2n\frac{s_n R}{r_n + R} = 2n\frac{s_{2n} R}{r_{2n}} = 2n S_{2n} = P_{2n}.$$

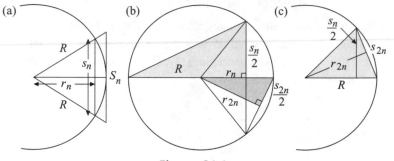

Figure S4.1.

Computing the area of the shaded triangle in Figure S4.1c in two different ways yields $(1/2)R(s_n/2) = (1/2)s_{2n}r_{2n}$, or $Rs_n = 2s_{2n}r_{2n}$. Thus

$$\sqrt{p_n P_{2n}} = n\sqrt{2s_n S_{2n}} = n\sqrt{2Rs_n s_{2n}/r_{2n}} = 2ns_{2n} = p_{2n}.$$

4.4 The problem is with the two valley folds at the 45° angle at the vertex. The two angles adjacent to it are both right angles, and with valley creases on both sides of the 45° angle, the two 90° angles will cover the 45° angle on the same side of the paper. When the creases are then pressed flat, the two 90° angles will intersect each other, and self-intersections of the paper are not allowed (in three-dimensional space) [Hull, 2004].

4.5 Since $p_n = 2nr \sin \pi/n$, we have $\lim_{n\to\infty}(n/\pi) \sin \pi/n = 1$. If we
let $\pi/n = \theta$, then $\theta \to 0^+$ as $n \to \infty$ and conversely, so that $\lim_{\theta\to 0+}$
$\sin\theta/\theta = 1$ [Knebelman, 1943].

4.6 Each of the p vertex angles measures $[1 - 2(q/p)]180°$, and hence the
sum is $(p - 2q)180°$.

4.7 From Corollary 4.2, the sum of the squares of the diagonals from a
given vertex and the two adjacent sides is $2nR^2$. Repeating for each
vertex counts squares of all the diagonals and all the sides twice. Hence
the sum in question is $(1/2)n(2nR^2) = n^2R^2$ [Ouellette and Bennett,
1979].

4.8 The first inequality is immediate from part (i) of Theorem 4.10 while
the second follows from the first by induction with the basis step
$T_3 = 1$.

4.9 The two acute angles at the base each measure $\pi/2 - \pi/n$, and each
of the other $n - 2$ obtuse angles measures $\pi - \pi/n$. See Figure 4.18,
and observe that at each vertex of the polygonal cycloid the angle in the
shaded triangle measures π/n and each of the two angles in the white
triangle measures $\pi/2 - \pi/n$.

4.10 We compute the area of the n-gon in two ways using the point Q and
the incenter I to triangulate the n-gon (see Figure S4.2). Let d_i denote
the perpendicular distances from Q to the sides and s the side length
of the n-gon. Then in Figure S4.2a, the area of the n-gon is given by
$s \sum d_i/2$ whereas in Figure 4.2b, the area is $nrs/2$. Hence $\sum d_i = nr$
as claimed.

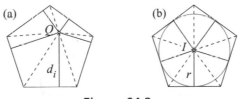

Figure S4.2.

4.11 (a) See Figure S4.3a. (b) They are concurrent. In the pentagon $V_1V_2V_3$
V_4V_{12} the angle sum is 540° while the angles at V_1, V_2, and V_3 are
each 150°, hence the angles at V_{12} and V_4 are each 45°. Thus all
three diagonals pass through the center of the shaded square. See
Figure S4.3b.

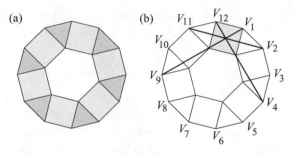

Figure S4.3.

Chapter 5

5.1 No. Let a, b, c denote the sides of a triangle, and assume $\sqrt{a} + \sqrt{b} = \sqrt{c}$. Because $\sqrt{a}$, $\sqrt{b}$, and $\sqrt{a+b}$ are the sides of a a right triangle by the Pythagorean theorem, we have $\sqrt{a+b} < \sqrt{a} + \sqrt{b}$, so that $\sqrt{a+b} < \sqrt{c}$. Thus we have $a + b < c$, which is impossible.

5.2 The equations in (5.1) are equivalent to $ab = 2rs$ and $s = c + r$, thus $(a - r)(b - r) = ab - (a + b)r + r^2 = 2rs - (2s - c)r + r^2 = r(c + r) = rs = K$.

5.3 The square in the triangle on the left has area equal to $2/4 = 1/2$ of the triangle, while the one on the right has area equal to $4/9$ of the triangle (see Figure S5.1), so the one on the left is larger [DeTemple and Harold, 1996].

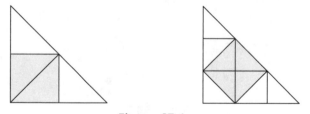

Figure S5.1.

5.4 See Figure 5.18. We prove $U/u = a/c$. Applying the law of sines to triangle ABB' yields $\sin \beta / \sin \angle BB'A = u/c$, and to triangle CBB' yields $\sin \beta / \sin \angle BB'C = U/a$. But $\sin \angle BB'A = \sin \angle BB'C$, hence $u/c = U/a$, or $U/u = a/c$.

5.5 (a) By Lemma 5.3, $abc = 4KR$ and thus

$$h_a + h_b + h_c = 2K\left[\frac{1}{a} + \frac{1}{b} + \frac{1}{c}\right] = \frac{abc}{2R}\left[\frac{ab + bc + ca}{abc}\right]$$

$$= \frac{ab + bc + ca}{2R}.$$

(b) Using Lemma 5.1,

$$\frac{1}{h_a} + \frac{1}{h_b} + \frac{1}{h_c} = \frac{a}{2K} + \frac{b}{2K} + \frac{c}{2K} = \frac{2s}{2K} = \frac{1}{r}.$$

5.6 (a) For convenience let the side of the square be 4 (see Figure S5.2a). Then the radius of the semicircle is 2, so that the striped right triangle has legs 2 and 4. But the dark gray right triangle is similar to the striped one, so its legs are 1 and 2, thus the light gray triangle is the 3:4:5 right triangle.

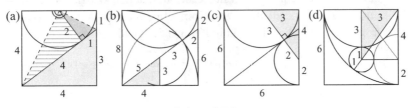

Figure S5.2.

(b) Now let the side of the square be 8 (see Figure S5.2b), and begin with the figure from part (a). Draw the quarter circle of radius 8 and center at the lower left corner of the square, mark off the 3:4:5 gray triangle, and draw a circle of radius 3 (only partially drawn in the figure) with the upper vertex of the gray triangle as center. This circle will be tangent to the two quarter circles and the bottom edge of the square.

(c) In this case we let the side of the square be 6 (see Figure S5.2c) and again begin with the figure from part (a). Draw the line segment perpendicular to the tangent line at the point of tangency to create the shaded gray triangle similar to the one from part (a). Since its hypotenuse is 5 and the radius of the semicircle is 3, one can now draw the semicircle of radius 2 along the right side of the square. Finally, rotate the entire figure 180°.

(d) Again using a square of side 6, reorient the gray triangle from part (c) into the position shown in Figure S5.2d and draw a quarter circle with radius 6 and center the upper right corner of the square. Since the hypotenuse of the shaded triangle is 5, a circle of radius 1 can be drawn as shown with center at the lower vertex of the shaded triangle. Finally, draw the second quarter circle and rotate the entire figure 180°.

5.7 (a) The result follows from the law of cosines applied to the triangles with sides m_a, $a/2$, b and m_a, $a/2$, c and the fact that cosines of supplementary angles have the same magnitude but opposite sign.

(b) Since $m_a^2 - m_b^2 = 3(b^2 - a^2)/4$, $a \leq b$ implies $m_a \geq m_b$ and similarly $b \leq c$ implies $m_b \geq m_c$.

5.8 We apply the Erdős-Mordell theorem 5.11 in the case where O is the circumcenter with $x = y = z = R$. As noted in the paragraph following the proof of the Erdős-Mordell theorem, the three inequalities in Lemma 5.4 are equalities in this case, so that $aR = bw + cv$, $bR = aw + cu$, and $cR = av + bu$. Computing the area K of the triangle from Figure 5.16b and invoking Lemma 5.1 yields

$$K = (au + bv + cw)/2 = rs = r(a + b + c)/2,$$

and thus

$$
\begin{aligned}
(a + b + c)(u + v + w) &= (au + bv + cw) + (bw + cv) \\
&\quad + (aw + cu) + (av + bu) \\
&= r(a + b + c) + R(a + b + c) \\
&= (a + b + c)(r + R),
\end{aligned}
$$

whence $u + v + w = r + R$.

5.9 When the triangle is equilateral, there is nothing to prove as the centers G, H, and O coincide. When the triangle is not equilateral, see Figure S5.3. Since CH is parallel to OM_c, the shaded triangles are similar, $|CG| = 2|GM_c|$, and the result follows.

5.10 We can reverse the steps from the solution to the pervious Challenge. If the centroid G and the circumcenter O coincide, then the medians are perpendicular to the sides, so each median is an altitude. If G and O are

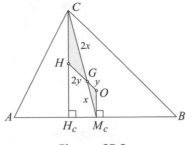

Figure S5.3.

distinct, draw the line GO and locate point H so that $|GH| = 2|GO|$ with G between H and O. Then the shaded triangles are similar, so that CH (and thus CH_c) is parallel to OM_c. But OM_c is perpendicular to AB, and hence so is CH_c. Similarly one can show that AH and BH (extended) are perpendicular to BC and AC, respectively.

5.11 Yes, as a consequence of the facts that angles inscribed in semicircles are right angles and the three altitudes of a triangle are concurrent. See Figure S5.4.

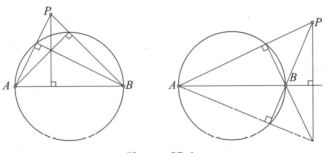

Figure S5.4.

5.12 See Figure S5.5a, where $a > b > c$. Since the sides are in arithmetic progression, $a + c = 2b$ so the semiperimeter is $s = 3b/2$. Hence the area K satisfies $K = rs = 3br/2$ and $K = bh/2$, thus $h = 3r$ so that the incenter I lies on the dashed line parallel to AC one-third of the way from AC to B (see Figure S5.5b). But the centroid G lies on the median BM_b one-third of the way from AC to B, hence G also lies on the dashed line [Honsberger, 1978].

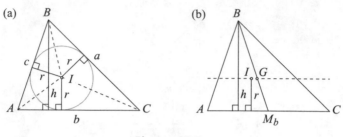

Figure S5.5.

5.13 Without loss of generality let the altitude to the hypotenuse have length 1 as shown in Figure S5.6.

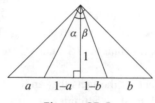

Figure S5.6.

Then $c = 2 - a - b$ and

$$\tan(\alpha + \beta) = \frac{(1 - a) + (1 - b)}{1 - (1 - a)(1 - b)} = \frac{2 - a - b}{a + b - ab} = \frac{c}{a + b - ab}.$$

Thus $\alpha + \beta = 45°$ if and only if $c = a + b - ab$. But $c^2 = a^2 + b^2$ if and only if $(2 - a - b)^2 = a^2 + b^2$, which simplifies to $2 - a - b = a + b - ab$ [Mortici, 2009].

5.14 (a) To show that Heron's formula implies the Pythagorean theorem, consider the isosceles triangle in Figure S5.7a composed of two right triangles with legs a and b, hypotenuse c, and area $K = ab/2$. The semiperimeter of the isosceles triangle is $c + b$, and hence its area $2K$ satisfies

$$2K = ab = \sqrt{(c + b) \cdot b \cdot b \cdot (c - b)} = b\sqrt{c^2 - b^2},$$

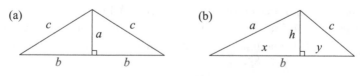

Figure S5.7.

and thus $a^2 = c^2 - b^2$. (This proof is from [Loomis, 1969], who attributes it to J. J. Posthumus.)

(b) To show that the Pythagorean theorem implies Heron's formula, consider the triangle in Figure S5.7b, where h is the altitude to side $b = x + y$ and $s = (a + b + c)/2$. Since its area $K = bh/2$, we have

$$16K^2 = 4b^2h^2 = 4b^2(a^2 - x^2) = (2ab)^2 - (2bx)^2.$$

But

$$2bx = b^2 + x^2 - (b - x)^2 = b^2 + x^2 - y^2$$
$$= b^2 + (x^2 + h^2) - (y^2 + h^2) = b^2 + a^2 - c^2.$$

Thus

$$16K^2 = (2ab)^2 - (a^2 + b^2 - c^2)^2$$
$$= [(a + b)^2 - c^2][c^2 - (a - b)^2]$$
$$= (a + b + c)(a + b - c)(a - b + c)(-a + b + c)$$
$$= 16s(s - a)(s - b)(s - c).$$

Chapter 6

6.1 From the law of cosines, we have $c^2 = a^2 + b^2 - 2ab \cos C$. Since $T = ab \sin C /2$, $2ab = 4T/\sin C$ and thus $c^2 = a^2 + b^2 - 4T \cot C$. Multiplication by $\sqrt{3}/4$ yields the desired result since $T_s = s^2\sqrt{3}/4$.

6.2 See (a) Figure S6.1 and (b) Figure S6.2.

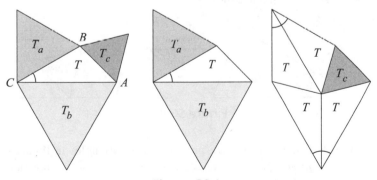

Figure S6.1.

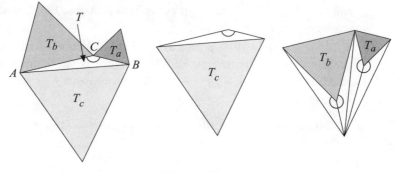

Figure S6.2.

6.3 In Figure 6.17 we have extended BC to C' so that $C'B = AB$ and drawn triangle $C'P'B$ congruent to triangle APB. Thus $C'P' = AP$ and $\angle PBP' = \angle ABC' \leq 60°$. Hence $BP \geq P'P$ and

$$AP + BP + CP \geq C'P' + P'P + PC > C'C = AB + BC.$$

Hence the sum of the distances to the vertices is a minimum when P coincides with B [Niven, 1981].

6.4 Without loss of generality assume $a \geq b \geq c$, and let x, y, z denote the lengths of the perpendicular segments from P to sides a, b, c, respectively. If h_a denotes the altitude to side a, then computing twice the area of the triangle in two different ways yields $ah_a = ax + by + cz$. But if the triangle is not equilateral, then either $a > b \geq c$ or $a \geq b > c$; in either case we have $ah_a < a(x + y + z)$ or $h_a < x + y + z$. Thus P lies at vertex A (and B if $a = b$).

6.5 See Figure S6.3 [Kung, 2002].

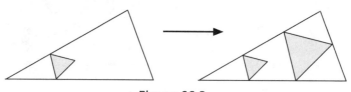

Figure S6.3.

6.6 The side lengths of the dark gray and white triangles in Figure 6.18 are $2n - m$ and $2m - 3n$, respectively, so that using the notation introduced in Section 6.1 to express the area of the equilateral triangles, we have $T_m = 3T_n + T_{2m-3n} - 3T_{2n-m}$. If we assume that $\sqrt{3}$ is rational, then $\sqrt{3} = m/n$ in lowest terms, so that $m^2 = 3n^2$, or equivalently, $T_m = 3T_n$. Hence $T_{2m-3n} = 3T_{2n-m}$, which implies that $\sqrt{3} = (2m-3n)/(2n-m)$, a contradiction since $0 < 2m - 3n < m$ and $0 < 2n - m < n$.

6.7 See Figure S6.4 [Andreescu and Gelca, 2000]. As in the proof of Theorem 6.5, we rotate ABC and the three segments to the positions shown in Figure 6.20b. Again triangle CPP' is equilateral. Then $|BP| = |AP'|$ and $|CP| = |PP'|$, hence the lengths of the sides of triangle APP' equal $|AP|, |BP|,$ ands $|CP|$. The degenerate case occurs when P lies on the circumcircle of triangle ABC, and van Schooten's theorem 6.5 holds.

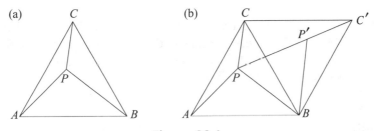

Figure S6.4.

6.8 Since the point E is a vertex shared by four triangles in the square, we locate it by drawing quarter circles with centers at A and B as shown in Figure S6.5a. We claim the base angles of the dark gray triangle are each 15°. Since triangle ABE is equilateral, $|AD| = |AE|, |BC| = |BE|$, and the light gray triangles are isosceles with apex angles equal to 30°. Hence their base angles equal 75°, and so the base angles of the dark gray triangle are each 15°.

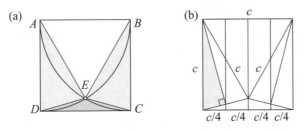

Figure S6.5.

6.9 See Figure S6.5b [Pinter, 1988].

6.10 (i) Let c denote the length of the side of the equilateral triangle. Then the area of each dark gray triangle is $c^2/8$ (from the preceding Challenge) and the area of the light gray triangle is $c^2/4$.

(ii) See Figure S6.6.

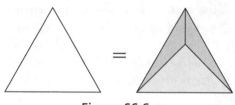

Figure S6.6.

6.11 Triangle AEF (see Figure 6.21) is an isosceles right triangle, hence $\angle AEF = 45°$. Similarly triangle BCE is isosceles with $\angle CBE = 30°$, hence $\angle BEC = 75°$. Thus the three angles at E below the dashed line sum to $180°$.

Chapter 7

7.1 See Figure S7.1.

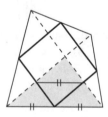

Figure S7.1.

7.2 Use (7.3), $\cos(\theta + \phi) = 2\cos^2((\theta + \phi)/2) - 1$, and the algebra in the paragraph preceding Corollary 7.1. Either pair of opposite angles can be used, since their half-sums are supplementary.

7.3 Set $d = 0$ in Corollary 7.1.

7.4 (a) Apply the theorem to a rectangle with sides a and b and diagonals c.

(b) See Figure 6.14a, and let s denote the side of the equilateral triangle. Then $s \cdot |AP| = s \cdot |BP| + s \cdot |CP|$, and hence $|AP| = |BP| + |CP|$.

(c) Let x denote the length of the diagonal, and consider the cyclic quadrilateral in the pentagon with sides 1, 1, 1, x and diagonals x and x. Then $x^2 = x \cdot 1 + 1 \cdot 1$, so x is the (positive) root of $x^2 = x + 1$, i.e., $x = \varphi$.

(d) Choose a cyclic quadrilateral one of whose diagonals is the diameter of a circle with diameter 1, and use the fact that in such a circle the length of any chord is equal to the sine of the angle it subtends.

7.5 See Figure S7.2. The area of Q is $K = p(h_1 + h_2)/2$, but $h_1 + h_2 \le q$, hence $K \le pq/2$. Equality holds if and only if $h_1 + h_2 \le q$, i.e., when the diagonals are perpendicular.

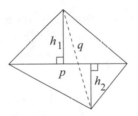

Figure S7.2.

7.6 See Figure 7.15, and observe that

$$a + b = 180° - (x + y) + 180° - (z + t)$$
$$= 360° - (x + y + z + t) = 360° - 180° = 180°.$$

7.7 (a) Each diagonal of Q splits Q into two triangles, as illustrated in Figure 7.8 (but here we do not assume that Q is cyclic). If the angle between sides a and b is θ then $K_1 = ab \sin \theta /2 \le ab/2$, and similarly for K_2, K_3, and K_4. Thus $K \le (ab + cd)/2$ and $K \le (ad + bc)/2$, with equality in one when two opposite angles are each 90°.

(b) From (a), $2K \le (ab + cd + ad + bc)/2 = (a + c)(b + d)/2$, with equality when all four angles are 90°.

(c) By the AM-GM inequality, $(a + c)(b + d) \le ((a + b + c + d)/2)^2 = L^2/4$, so from (b), $K \le L^2/16$. A rectangle with $a + c = b + d$ is a square.

7.8 Let A', B', C', D' denote the centers of the circles, as shown in Figure S7.3. Since A' and B' are the same distance r from the line AB, $A'B'$ is parallel to AB. Similarly $A'D'$ is parallel to AD, and thus the angles at A and A' are equal. Similarly for the other pairs B and B', etc.

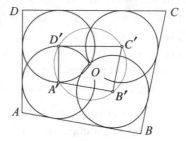

Figure S7.3.

Since each circle center is a distance r from the point O of concurrency, they all lie on a circle of radius r with center O, and thus $A'B'C'D'$ is cyclic. Hence A' and C' are supplementary as are A and C, so that $ABCD$ is also cyclic.

7.9 In this case, P can be *any* point in the parallelogram! See Figure S7.4, where we have colored triangles with equal areas the same. Each pair of opposite triangles is composed of four differently colored small triangles.

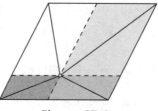

Figure S7.4.

7.10 Let P be the second point P of intersection of the circles ARS and BQS (see Figure S7.5 for the case when P lies inside the triangle). Since $ARPS$ is a cyclic quadrilateral, $x = y$, and since $BQPS$ is cyclic, $y = z$. Hence $x = z$ and $CQPR$ is also cyclic. When P lies outside the triangle, the theorem remains true but the argument is different [Honsberger, 1995].

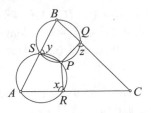

Figure S7.5.

7.11 Let P be an arbitrary point in or on the convex quadrilateral $ABCD$, and let Q denote the intersection of the two diagonals, as illustrated in Figure S7.6.

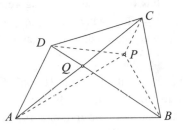

Figure S7.6.

Then $QA + QB + QC + QD = AC + BD \le PA + PB + PC + PD$.

Chapter 8

8.1 The intersection of $y = (1/\sqrt{n})x$ with $x^2 + y^2 = 1$ yields $ny^2 + y^2 = 1$, so that $y = 1/\sqrt{n+1}$.

8.2 The smaller square has (a) $4/10 = 2/5$ and (b) $1/13$ the area of the larger square. See Figure S8.1 [DeTemple and Harold, 1996].

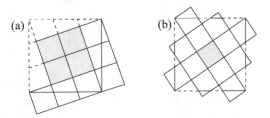

Figure S8.1.

8.3 If M is odd, then $M = 2k + 1 = (k+1)^2 - k^2$ and if M is a multiple of 4, then $M = 4k = (k+1)^2 - (k-1)^2$. Conversely, if $M = A^2 - B^2$, then M is congruent to 0, 1, or 3 (mod 4) since A^2 and B^2 are each congruent to 0 or 1 (mod 4).

8.4 If $2M = a^2 + b^2$, then a and b are either both even or both odd. In either case $(a+b)/2$ and $(a-b)/2$ are both integers, so that

$$M = \left(\frac{a+b}{2}\right)^2 + \left(\frac{a-b}{2}\right)^2.$$

8.5 If $\sqrt{2}$ were rational, say $\sqrt{2} = c/a$ (a and c integers) in lowest terms, then $c^2 = 2a^2$. Thus (a, a, c) is a Pythagorean triple, so by Theorem 8.8 we have $n^2 = 2m^2$ where $n = 2a - c$ and $m = c - a$. So $\sqrt{2} = (2a - c)/(c - a)$, contradicting the assumption that $\sqrt{2} = c/a$ in lowest terms.

8.6 Let K denote the area of the triangle. Heron's formula (see Theorem 5.7) is equivalent to $16K^2 = 2(a^2b^2 + b^2c^2 + c^2a^2) - (a^4 + b^4 + c^4)$, which expresses the square of the area of a triangle in terms of the areas of the squares on the sides. With $a^2 = 370$, $b^2 = 116$, and $c^2 = 74$ we have $16K^2 = 1936$, and thus $K = 11$ acres.

8.7 The inscribed square has one-half the area of the circumscribed square. See Figure S8.2.

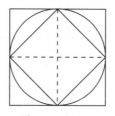

Figure S8.2.

8.8 Yes: $3(a^2 + b^2 + c^2) \equiv (a+b+c)^2 + (b-c)^2 + (c-a)^2 + (a-b)^2$.

8.9 See Figure S8.3.

8.10 Duplicate the square and rotate the copy 90° counterclockwise and place it immediately to the left of the given square, as illustrated in Figure S8.4. Since the two segments of length 2 are perpendicular, they are the legs of an isosceles right triangle whose hypotenuse QP has

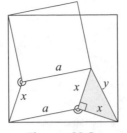

Figure S8.3.

length $2\sqrt{2}$. Hence the segments of length 1, $2\sqrt{2}$, and 3 also form a right triangle, and thus the angle at P between the segments of lengths 1 and 2 is $45° + 90° = 135°$.

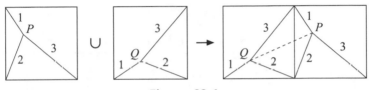

Figure S8.4.

Chapter 9

9.1 (a) We first find the polar equation of the cissoid. The circle C is given by $r = 2a \sin \theta$, the line T by $r = 2a \csc \theta$, and the line OP by $\theta = t$ for some t in $(0, \pi)$. Thus the polar coordinates of A and B are $(2a \sin t, t)$ and $(2a \csc t, t)$, respectively, and $|AB| = 2a(\csc t - \sin t)$. Setting $|OP| = r$ and $t = \theta$ yields the polar equation of the cissoid: $r = 2a(\csc \theta - \sin \theta)$ which in Cartesian coordinates is $x^2 = y^3/(2a - y)$.

(b) To represent $\sqrt[3]{2}$, draw the line $x + 2y = 4a$ through the points $(4a,0)$ and $(0,2a)$, and let it intersect the cissoid at Q. See Figure S9.1. The slope of OQ is $1/\sqrt[3]{2}$, and OQ extended intersects T at $R = (2a\sqrt[3]{2}, 2a)$. Thus the distance from the point of tangency of T with C to R is $2a\sqrt[3]{2}$.

9.2 One parabola has focus $(0, 1/4)$ and directrix $y = -1/4$ (i.e., $y = x^2$) and the other has focus $(1/2, 0)$ and directrix $x = -1/2$ (i.e., $2x = y^2$). Hence $2x = (x^2)^2$, or $x = \sqrt[3]{2}$.

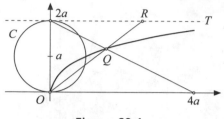

Figure S9.1.

9.3 See the comment in the paragraph following the proof of Theorem 9.3.

9.4 See Figure 9.2. (a) The shaded portion of the circle is equal in area to the shaded hexagon, which is squarable. (b) The object can be cut into three pieces that form a square [Gardner, 1995].

(a)   (b)

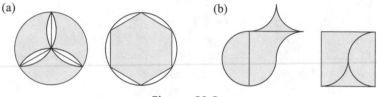

Figure S9.2.

9.5 Let R denote the radius of the original circle and S, C, and L the areas of the circular segment to the left of the dashed chord, the white circle, and Leonardo's claw, respectively. Then $S = \pi R^2/4 - R^2/2$ and the radius of the white circle is $R\sqrt{2}/2$. Thus $L = \pi R^2 - C - 2S = R^2$, which is the area of the light gray square in the claw's grasp.

9.6 Our proof is one attributed to G. Patruno in [Honsberger, 1991]. See Figure S9.3, where we have drawn AM, MB, and MC, and have located D on AB so that $|AD| = |BC|$. Now $\angle MAB = \angle MCB$ and $|AM| = |MC|$, so that $\triangle ADM$ is congruent to $\triangle CBM$ and hence

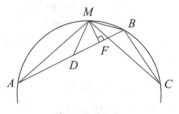

Figure S9.3.

$|MD| = |MB|$. Thus $\triangle MDB$ is isosceles, so that MF is both the altitude and the median from M, and $|DF| = |FB|$. Hence $|AF| = |AD| + |DF| = |BC| + |FB|$ as desired.

9.7 Let $t \in [0, 1]$, and let (x, y) be the coordinates of the point of intersection of OA' and $A'B'$ in Figure 9.17a. Then x and y are given parametrically by $y = 1 - t$ and $x = y \tan(\pi t/2)$ for $t \in [0, 1)$. Eliminating the parameter t yields $x = y \cot(\pi y/2)$ for $y \in (0, 1]$. Taking the limit of x as $y \to 0^+$ yields $x = 2/\pi$ for $y = 0$.

9.8 See Figure S9.4.

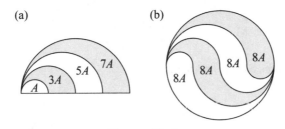

(a)   (b)

Figure S9.4.

9.9 Since the area under Γ is constant, we need only minimize the area of the trapezoid whose upper edge is the tangent line. The area of a trapezoid is equal to its base times its height at the midpoint c of the base, so it suffices to minimize the height to the point Q in Figure S9.5. This occurs when $Q = P$, that is, the tangent at P is used [Paré, 1995].

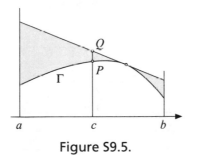

Figure S9.5.

Chapter 10

10.1 Let a and b denote the legs and c the hypotenuse of the triangle. Then in Figure S10.1a we see four copies of the triangle and the small white square in a square with area c^2 and in Figure S10.1b we see the same figures in two squares with areas a^2 and b^2. This proof is usually attributed to the Indian mathematician Bhāskara (1114–1185), who labeled it with only the word *Behold!*

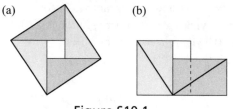

(a)   (b)

Figure S10.1.

10.2 See Figures S10.2a and S10.2b.

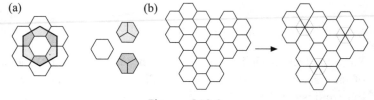

(a)   (b)

Figure S10.2.

10.3 Two copies of the pentagon can be joined to form a hexagon that tiles the plane. See Figure S10.3.

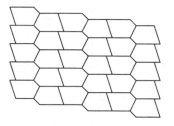

Figure S10.3.

10.4 (a) and (b) see Figure S10.4, (c) and (d) see Figure S10.5.

(a)   (b)

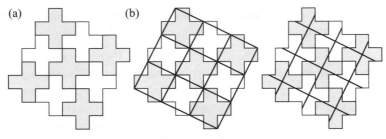

Figure S10.4.

(c)   (d)

Figure S10.5.

10.5 See Figure S10.6.

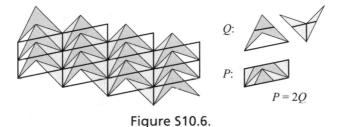

Q:

P:

$P = 2Q$

Figure S10.6.

10.6 All have translation symmetry. Additional symmetries are (a) rotation,
(b) rotation, vertical and glide reflections, (c) horizontal and glide reflections, (d) all four, (e) vertical reflection, (f), glide reflection, and
(g) none.

10.7 **Theorem** [Singmaster, 1975]. *Consider a deficient 1-by-2k chessboard
where the squares are alternately colored with two colors, numbered
$1, 2, \ldots, 2k$ from left to right, and squares i and j ($i < j$) have been*

removed. The chessboard can be tiled with dominoes if and only if the squares removed have different colors and i is odd.

Proof. Assume the deficient board can be tiled. Then the removed squares have different colors, and there must be an even number of squares to the left of square i, thus i must be odd. Now assume that the squares removed have different colors and i is odd. Then j is even, and there are an even number of squares to the left of i, between i and j, and to the right of j, so that the deficient board can be tiled.

10.8 It is impossible to tile any portion of the board that includes one edge of the board.

10.9 Since each tetromino consists of four squares, 4 divides mn, so either m or n must be even. Suppose m is even, and is the length of the horizontal side. Color the chessboard column-wise as illustrated in Figure 10.22a. Let x be the number of L-tetrominoes that cover one light gray and three dark gray squares and y the number of L-tetrominoes that cover one dark gray and three light gray squares. Then $x + 3y = y + 3x$ (i.e., $x = y$) and $x + y = mn/4$, thus $2x = mn/4$, so $mn = 8x$.

10.10　(a) We will show that the smaller 3×7 board where each square is colored black or white has a rectangle whose four corner squares all have the same color. Two similarly colored squares in the same column will be called a *doublet*. Since each column has at least one doublet, the 3×7 board contains at least seven doublets, and by the pigeonhole principle four of those (in four different columns) must be the same color, say black. Each doublet determines a pair of rows, but since there are only three different pairs of rows, two doublets (again by the pigeonhole principle) must determine the same pair of rows. The four black squares in these two doublets are the corners of the desired rectangle.

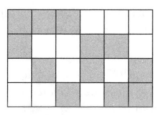

Figure S10.7.

(b) Since two black and two white squares can be arranged in a column six different ways, we can construct the following example of a 4×6 chessboard in which the four corner squares of every such rectangle described in the problem are not the same color [Beresin et al., 1989]:

10.11 Partition the hexagons into congruent isosceles triangles as shown in Figure S10.8 and count the number of triangles in each hexagon. Thus their areas are in the ratio 24:18:12, or 4:3:2.

Figure S10.8.

10.12 Two colors suffice. Take the folded flat origami and place it flat on a table. Color all regions blue if they face up (away from the table) and color all regions red if they face down (into the table). Neighboring regions of the crease pattern have a crease between them, and hence point in different directions when folded. Thus they are colored differently [Hull, 2004].

10.13 (a) The state of Nevada is bordered by five states: Oregon, Idaho, Utah, Arizona, and California (see Figure S10.9). This ring of five states cannot be colored with two colors, as two adjacent states would then have the same color. So three colors are required for the ring, and since Nevada borders on each of those states, it requires a fourth color. The same is true elsewhere on the map, as West Virginia is bordered by five states and Kentucky is bordered by seven states.

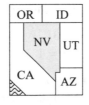

Figure S10.9.

(b) The nations of Belgium, Germany, France, and Luxembourg border one another in a manner similar to the map in Figure 10.38.

10.14 Yes, using the strategy proposed by O'Beirne in Section 10.7. See Figure S10.10.

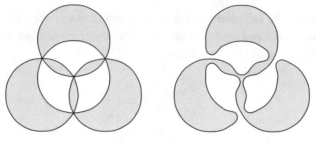

Figure S10.10.

Chapter 11

11.1 See Figure 11.1. Theorem 11.1 yields $4K^2 = b^2c^2 + a^2c^2 + a^2b^2$, or equivalently, $(2K/abc)^2 = (1/a)^2 + (1/b)^2 + (1/c)^2$. But the volume of the right tetrahedron is both $abc/6$ and $pK/3$, hence $2K/abc = 1/p$.

11.2 Assume there exists a polyhedron such that each face has a different number of sides. Consider the face with the largest number m of sides. Sharing the edges of this face are m other faces, each with a number of sides between three and $m - 1$, inclusive. By the pigeonhole principle, at least two of them must have the same number of sides, a contradiction.

11.3 (a) $2V - 2E + 2F = 4$ and $2E \geq 3F$ implies $2V - F \geq 4$, and $2F - V \geq 4$ is established similarly.

(b) $3V - 3E + 3F = 6$ and $2E \geq 3V$ implies $3F - E \geq 6$, and $3V - E \geq 6$ is established similarly.

11.4 If $E = 7$, then $3F \leq 14$ and $3V \leq 14$, and hence $F = V = 4$. But then $V - E + F = 4 - 7 - 4 = 1 \neq 2$.

11.5 The number of ways to assign F colors to F faces is $F!$, but many of the colorings coincide upon rotation. To compare colorings, we will now place the solid on, say, a table in standard position, by choosing one face (say red) in F ways to lie on the table and then another (say blue) to be facing the viewer, If each face is a n-gon, this can be done in n ways. Hence we divide $F!$ by nF, but for Platonic solids $nF = 2E$.

11.6 The obvious answer $5 + 4 - 2 = 7$ is incorrect; the correct answer is five. To see this, place two copies of the pyramid side by side, as illustrated in Figure S11.1. Since the distance between the tops of the pyramids is the same as the side of the square base, the space between the two pyramids is congruent to the tetrahedron, and hence after gluing the tetrahedron and pyramid together, the front face of the tetrahedron lies in the same plane as the two shaded faces of the pyramids. Hence the resulting polyhedron has five faces—the square base, two rhombus faces, and two equilateral triangles [Halmos, 1991].

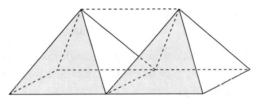

Figure S11.1.

11.7 First note that an isosceles tetrahedron can be inserted into a rectangular box as shown in Figure S11.2, with each of the six edges of the tetrahedron a diagonal of one of the six faces of the box.

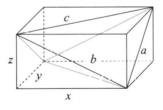

Figure S11.2.

If the dimensions of the box are x, y, z as shown in the figure, then

$$x^2 + y^2 = c^2, \quad y^2 + z^2 = a^2, \quad z^2 + x^2 = b^2$$

so that

$$x = \sqrt{\frac{b^2 + c^2 - a^2}{2}}, \quad y = \sqrt{\frac{c^2 + a^2 - b^2}{2}}, \quad \text{and}$$

$$z = \sqrt{\frac{a^2 + b^2 - c^2}{2}}.$$

The volume of the tetrahedron equals the volume of the box minus the volume of the four right tetrahedra with perpendicular sides of lengths x, y, z, i.e., $xyz - 4 \cdot (xyz/6) = xyz/3$. Hence the volume of the tetrahedron is $\sqrt{(a^2 + b^2 - c^2)(c^2 + a^2 - b^2)(b^2 + c^2 - a^2)/72}$ as claimed. The circumscribed sphere for the tetrahedron is the same as for the box, so its diameter equals the length of the diagonal of the box, and $\sqrt{x^2 + y^2 + z^2} = \sqrt{(a^2 + b^2 + c^2)/2}$ [Andreescu and Gelca, 2000].

Chapter 12

12.1 Since $\mathbb{Q}_+$ is denumerable, we can list its elements $\{q_1, q_2, q_3, \ldots\}$, and this list can be put into a one-to-one correspondence with the list $\{0, q_1, -q_1, q_2, -q_2, q_3, -q_3, \ldots\}$, which is Q.

12.2 Let S be the set of points in the plane with integer coordinates. For each θ in $[0, \pi)$, let $S_\theta \subset S$ be the points of S in an infinite strip of inclination θ and width greater than 1. There are nondenumerably many S_θ, each with infinitely many points, but the intersection of any two is finite since the intersection of any two strips is a bounded parallelogram [Buddenhagen, 1971].

12.3 (a) $\sum_{i=1}^{n} a_i \cdot \sum_{i=1}^{n} \dfrac{1}{a_i} \geq \left(n \sqrt[n]{a_1 a_2 \cdots a_n} \right) \left(n \sqrt[n]{\dfrac{1}{a_1} \dfrac{1}{a_2} \cdots \dfrac{1}{a_n}} \right) = n^2$,

   (b) $\sum_{i=1}^{n} \left(\sqrt{a_i}^2 \cdot \sum_{i=1}^{n} \left(\sqrt{1/a_i} \right)^2 \geq \left(\sum_{i=1}^{n} 1 \right)^2 = n^2$.

12.4 Let $x, y > 0$ and apply the Cauchy-Schwarz inequality to $\{\sqrt{x}, \sqrt{y}\}$ and $\{\sqrt{y}, \sqrt{x}\}$:

$$2\sqrt{xy} = \sqrt{x}\sqrt{y} + \sqrt{y}\sqrt{x} \leq \sqrt{x+y}\sqrt{x+y} = x + y.$$

12.5 We apply the AM-GM (12.2) inequality to a set of n numbers with $n-1$ 1s and one positive number y: $(n - 1 + y)/n \geq y^{1/n}$, or equivalently,

$$\left(1 + \frac{y-1}{n} \right)^n \geq y.$$

Since this holds for all n, we can take the limit as $n \to \infty$ to obtain $e^{y-1} \geq y$. Multiplying both side by e yields $e^y \geq ey$, and on setting $y = x/e$ we obtain $e^{x/e} \geq x$, equivalent to $e^{1/e} \geq x^{1/x}$.

12.6 If f satisfies (12.2), then $f(x+y) + f(x-y) = f(x) + f(y) + f(x) - f(y) = 2f(x)$. If $f(x+y) + f(x-y) = 2f(x)$, then setting $x = 0$ yields $f(y) + f(-y) = 0$. Since $f(y+x) + f(y-x) = 2f(y)$, addition yields $2f(x+y) = 2f(x) + 2f(y)$, i.e., f satisfies (12.2).

12.7 For any $x > 0$ there is a positive y such that $x = y^2$, hence $f(x) = [f(y)]^2 > 0$. Thus we can take logarithms of both sides of $f(xy) = f(x)f(y)$ to yield $\ln f(xy) = \ln f(x) + \ln f(y)$. So $\ln f(x)$ is a continuous solution to (12.7). Hence $\ln f(x) = k \ln x$, or $f(x) = x^k$ for some constant k.

12.8 Let a, b, c be the sides opposite angles A, B, C, respectively. Then $a = 2R \sin A$, $b = 2R \sin B$, and $c = 2R \sin C$. The concavity of the sine on $[0, \pi]$ now yields

$$a + b + c = 6R \left(\frac{\sin A + \sin B + \sin C}{3} \right)$$

$$\leq 6R \sin \left(\frac{A + B + C}{3} \right)$$

$$= 6R \sin 60°$$

$$= 3\sqrt{3}R.$$

12.9 (a) If $f(x) \leq x$, then $f(0) \leq 0$, which combines with Theorem 12.12(a) to give $f(0) = 0$. Replacing x by $-x$ in Theorem 12.12(d) yields $f(x) \geq -f(-x)$, and hence $f(x) \geq -f(-x) \geq -(-x) = x$, so that $f(x) = x$.

(b) No. $|x|$ is subadditive and $|x| \geq x$, but $|x| \neq x$ for negative x [Small, 2007].

12.10 Summing a geometric series and a telescoping series yields [Walker, 2002]

$$\sum_{k=2}^{\infty} [\zeta(k) - 1] = \sum_{k=2}^{\infty} \sum_{n=2}^{\infty} 1/n^k = \sum_{n=2}^{\infty} \sum_{k=2}^{\infty} 1/n^k$$

$$= \sum_{n=2}^{\infty} \frac{1/n^2}{1 - (1/n)} = \sum_{n=2}^{\infty} \frac{1}{n^2 - n}$$

$$= \sum_{n=2}^{\infty} \left(\frac{1}{n-1} - \frac{1}{n} \right) = 1.$$

12.11 Since $\displaystyle \binom{2n}{n} \frac{1}{2^{2n}} = \frac{1 \cdot 3 \cdot 5 \cdots (2n-1)}{2 \cdot 4 \cdot 6 \cdots (2n)}$, using (12.3) yields

$$\lim_{n \to \infty} \frac{1 \cdot 3 \cdots (2n-1)}{2 \cdot 4 \cdots (2n)} \sqrt{\pi n}$$

$$= \lim_{n \to \infty} \frac{1 \cdot 3 \cdots (2n-1)}{2 \cdot 4 \cdots (2n)} \sqrt{2n+1} \sqrt{\frac{\pi n}{2n+1}}$$

$$= \sqrt{\frac{2}{\pi}} \sqrt{\frac{\pi}{2}} = 1.$$

Alternatively, use Stirling's approximation to $(2n)!$ and $n!$ and simplify.

12.12 While Burnside's formula can be proved using approximations to the integral $\int_1^{n+1/2} \ln x \, dx$, the following proof [Keiper, 1979] is simpler. Using Stirling's formula, we have

$$n! \sim \sqrt{2\pi n} \cdot n^n e^{-n} = \sqrt{2\pi} \left(\frac{n}{n+1/2} \right)^{n+1/2} e^{-n} (n+1/2)^{n+1/2}.$$

However,

$$\left(\frac{n}{n+1/2} \right)^{n+1/2} = \left(1 - \frac{1/2}{n+1/2} \right)^{n+1/2}$$

which approaches $e^{-1/2}$ as $n \to \infty$, which completes the proof.

12.13 From the solution to Challenge 12.1, we have $1 \cdot 3 \cdot 5 \cdots (2n-1) \sim 2^n n! / \sqrt{\pi n}$. Now use Stirling's formula for $n!$.

References

What is now proved was once only imagined.
William Blake
The Marriage of Heaven and Hell

E. A. Abbott, *Flatland: A Romance of Many Dimensions*, Seeley & Co., London, 1884.

A. D. Abrams and M. J. Paris, The probability that $(a, b) = 1$, *College Mathematics Journal*, 23 (1992), p. 47.

J. Aczél, *Lectures on Functional Equations and their Applications*, Academic Press, New York, 1966.

J. Aczél and C. Alsina, Trisection of angles, classical curves and functional equations. *Mathematics Magazine,* 71 (1998), pp. 182–189.

J. Aczél and J. Dhombres, *Functional Equations in Several Variables*, Cambridge University Press, Cambridge, 1989.

M. Aigner and G. M. Ziegler, *Proofs from THE BOOK*, 2nd edition, Springer, Berlin, 2001.

C. Alsina and R. B. Nelsen, *Math Made Visual: Creating Images for Understanding Mathematics*, Mathematical Association of America, Washington, 2006.

———, A visual proof of the Erdös-Mordell inequality, *Forum Geometricorum*, 7 (2007), pp. 99–102.

———, Geometric proofs of the Weitzenböck and Hadwiger-Finsler inequalities, *Mathematics Magazine*, 81 (2008), 216–219.

———, *When Less is More: Visualizing Basic Inequalities*, Mathematical Association of America, Washington, 2009.

T. Andreescu and R. Gelca, *Mathematical Olympiad Challenges*, Birkhaüser. Boston, 2000.

T. Andreescu and Z. Feng, *USA and International Mathematical Olympiads 2001*, Mathematical Association of America, Washington, 2002.

T. M. Apostol, Irrationality of the square root of two—A geometric proof. *American Mathematical Monthly* 107 (2000), pp. 841–842.

T. M. Apostol and M. A. Mnatsakanian, Cycloidal areas without calculus, *Math Horizons*, September 1999, pp. 12–16.

A. Arcavi and A. Flores, Mathematics without words, *College Mathematics Journal*, 31 (2000), p. 392.

E. F. Assmus, Jr., Pi, *American Mathematical Monthly* 92 (1985), pp. 213–214.

L. Bankoff and C. W. Trigg, The ubiquitous 3:4:5 triangle, *Mathematics Magazine*, 47 (1974), pp. 61–70.

K. Bankov, Applications of the pigeon-hole principle. *Mathematical Gazette*, 79 (1995), pp. 286–292.

E. Beckenbach and R. Bellman, *An Introduction to Inequalities*, Mathematical Association of America, Washington, 1961.

s.-m. belcastro and T. C. Hull, Classifying frieze patterns without using groups, *College Mathematics Journal*, 33 (2002), pp. 93–98.

A. T. Benjamin and J. J. Quinn, *Proofs That Really Count*, Mathematical Association of America, Washington, 2003.

M. Beresin, E. Levine, J. Winn, A chessboard coloring problem, *College Mathematics Journal*, 20 (1989), pp. 106–114.

D. Blatner, *The Joy of π*, Walker and Co., New York, 1997.

D. M. Bloom, A one-sentence proof that $\sqrt{2}$ is irrational. *Mathematics Magazine* 68 (1995), p. 286.

A. Bogomolny, Three Circles and Common Chords from *Interactive Mathematics Miscellany and Puzzles* http://www.cut-the-knot.org/proofs/circlesAnd-Spheres.shtml, Accessed 11 September 2009.

H. C. Bradley, Solution to problem 3028, *American Mathematical Monthly*, 37 (1930), pp. 158–159.

J. R. Buddenhagen, Subsets of a countable set, *American Mathematical Monthly*, 78 (1971), pp. 536–537.

J. A. Bullard, Properties of parabolas inscribed in a triangle, *American Mathematical Monthly*, 42 (1935), pp. 606–610.

———, Further properties of parabolas inscribed in a triangle, *American Mathematical Monthly*, 44 (1937), pp. 368–371.

P. S. Bullen, D. S. Mitrinovic, and P. M. Vasic, *Means and Their Inequalities*, Kluwer Academic Publishers, Dordrecht, 1988.

S. L. Campbell, Countability of sets, *American Mathematical Monthly*, 93 (1986), pp. 480–481.

L. Carroll, *Pillow Problems and a Tangled Tale,* Dover Publications, New York, 1958.

M. Chamberland, The series for e via integration. *College Mathematics Journal* 30 (1999), p. 397.

R. Chapman, Evaluating $\zeta(2)$. www.secamlocal.ex.ac.uk/people/staff/rchapma/etc/zeta2.pdf, 2003.

V. Chvátal, A combinatorial theorem in plane geometry, *J. Combinatorial Theory, Ser. B*, 18 (1975), pp. 39–41.

A. J. Coleman, A simple proof of Stirling's formula, *American Mathematical Monthly*, 58 (1951), pp. 334–336.

J. H. Conway and R. R. Guy, *The Book of Numbers*, Springer-Verlag, New York, 1996.

H. S. M. Coxeter, A problem of collinear points. *American Mathematical Monthly*, 55 (1948), pp. 26–28.

———, *Introduction to Geometry*, John Wiley & Sons, New York, 1961.

H. S. M. Coxeter and S. L. Greitzer, *Geometry Revisited*, Mathematical Association of America, Washington, 1967.

P. R. Cromwell, *Polyhedra*, Cambridge University Press, Cambridge, 1997.

A. Cupillari, Proof without words, *Mathematics Magazine*, 62 (1989), p. 259.

CUPM, *Undergraduate Programs and Courses in the Mathematical Sciences: CUPM Curriculum Guide 2004*, Mathematical Association of America, Washington, 2004.

——, *CUPM Discussion Papers about Mathematics and the Mathematical Sciences in 2010: What Should Students Know?* Mathematical Association of America, Washington, 2001.

G. David and C. Tomei, The problem of the calissons, *American Mathematical Monthly*, 96 (1989), pp. 429–431.

M. de Guzmán, *Cuentos con Cuentas*, Red Olímpica, Buenos Aires, 1997.

E. D. Demaine and J. O'Rourke, *Geometric Folding Algorithms: Linkages, Origami, Polyhedra*, Cambridge University Press, New York, 2007.

D. DeTemple and S. Harold, A round-up of square problems, *Mathematics Magazine*, 69 (1996), pp. 15–27.

N. Do, Art gallery theorems, *Gazette of the Australian Mathematical Society*, 31 (2004), pp. 288–294.

H. Dörrie, *100 Great Problems of Elementary Mathematics*, Dover Publications, Inc., New York, 1965.

W. Dunham, *Euler, The Master of Us All*, Mathematical Association of America, Washington, 1999.

R. A. Dunlap, *The Golden Ratio and Fibonacci Numbers*, World Scientific, Singapore, 1997.

R. Eddy, A theorem about right triangles, *College Mathematics Journal*, 22 (1991), p. 420.

A. Engel, *Problem-Solving Strategies*, Springer, New York, 1998.

D. Eppstein, Nineteen proofs of Euler's formula: $V - E + F = 2$, *The Geometry Junkyard*, http://www.ics.uci.edu/~eppstein/junkyard/euler, 2005.

P. Erdős, Problem 3740, *American Mathematical Monthly*, 42 (1935), p. 396.

——, Problem 4064. *American Mathematical Monthly*, 50 (1943), p. 65.

M. A. Esteban, *Problemas de Geometría*, FESPM, Badajoz, 2004.

L. Euler, in: *Leonhard Euler und Christian Goldbach, Briefwechsel 1729–1764*, A.P. Juskevic and E. Winter (editors), Akademie Verlag, Berlin, 1965.

H. Eves, *In Mathematical Circles*, Prindle, Weber & Schmidt, Inc., Boston, 1969.

―――, *An Introduction to the History of Mathematics, Fifth Edition*, Saunders College Publishing, Philadelphia, 1983.

S. Fisk, A short proof of Chvátal's watchman theorem, *J. Combinatorial Theory, Ser. B*, 24 (1978), p. 374.

D. Flannery, *The Square Root of Two*, Copernicus, New York, 2006.

G. Frederickson, *Dissections: Plane and Fancy*, Cambridge University Press, New York, 1997.

K. Fusimi, Trisection of angle by Abe, *Saiensu* supplement, (October 1980), p. 8.

J. W. Freeman, The number of regions determined by a convex polygon, *Mathematics Magazine*, 49 (1976), pp. 23 25.

M. Gardner, *More Mathematical Puzzles and Diversions*, Penguin Books, Harmondsworth, England, 1961.

―――, Mathematical Games, *Scientific American*, November 1962, p. 162.

―――, *Sixth Book of Mathematical Games from Scientific American,* Freeman, San Francisco, 1971.

―――, Mathematical games, *Scientific American*, October 1973, p. 115.

―――, *Time Travel and Other Mathematical Bewilderments*, Freeman, New York, 1988.

―――, *Mathematical Carnival*, Mathematical Association of America, Washington, 1989.

―――, *New Mathematical Diversions*, revised edition. Mathematical Association of America, Washington, 1995.

D. Goldberg, personal communication.

S. Golomb, Checker boards and polyominoes, *American Mathematical Monthly*, 61 (1954), pp. 675–682.

————, *Polyominoes*, Charles Scribner's Sons, New York, 1965.

J. Gomez, Proof without words: Pythagorean triples and factorizations of even squares, *Mathematics Magazine*, 78 (2005), p. 14.

N. T. Gridgeman, Geometric probability and the number π, *Scripta Mathematica*, 25 (1960), pp. 183–195.

C. M. Grinstead and J. L. Snell, *Introduction to Probability, Second Revised Edition*, American Mathematical Society, Providence, 1997.

D. Gronau, A remark on Sincov's functional equation, *Notices of the South African Mathematical Society*, 31 (2000), pp. 1–8.

B. Grünbaum, Polygons, in *The Geometry of Metric and Linear Spaces*, L. M. Kelly, editor, Springer-Verlag, New York, 1975, pp. 147–184.

B. Grünbaum and G. C. Shephard, *Tilings and Patterns*, W. H. Freeman, New York, 1987.

A. Gutierrez, *Geometry Step-by-Step from the Land of the Incas*. www.agutie.com.

R. Guy, There are three times as many obtuse-angled triangles as there are acute-angled ones, *Mathematics Magazine*, 66 (1993), pp. 175–179.

M. Hajja, A short trigonometric proof of the Steiner-Lehmus theorem, *Forum Geometricorum*, 8 (2008a), pp. 39–42.

————, A condition for a circumcriptable quadrilateral to be cyclic, *Forum Geometricorum*, 8 (2008b), pp. 103–106.

P. Halmos, *Problems for Mathematicians Young and Old*, Mathematical Association of America, Washington, 1991.

J. Hambidge, *The Elements of Dynamic Symmetry*, Dover Publications, Inc., New York, 1967.

G. Hanna and M. de Villiers, ICMI Study 19: Proof and proving in mathematics education. *ZDM Mathematics Education*, 40 (2008), pp. 329–336.

G. H. Hardy, *A Mathematician's Apology*, Cambridge University Press, Cambridge, 1969.

G. H. Hardy and E. M. Wright, *An Introduction to the Theory of Numbers, Fourth Edition*, Oxford University Press, London, 1960.

J. D. Harper, The golden ratio is less than $\pi^2/6$. *Mathematics Magazine*, 69 (1996), p. 266.

———, Another simple proof of $1 + 1/2^2 + 1/3^2 + \cdots = \pi^2/6$, *American Mathematical Monthly* 110 (2003), pp. 540–541.

K. Hatori, http://origami.ousaan.com/library/conste.html, 2009.

J. Havil, *Gamma: Exploring Euler's Constant*, Princeton University Press, Princeton, 2003.

R. Herz-Fischler, A "very pleasant theorem," *College Mathematics Journal*, 24 (1993), pp. 318–324.

L. Hoehn, A neglected Pythagorean-like formula, *Mathematical Gazette*, 84 (2000), pp. 71–73.

P. Hoffman, *The Man Who Loved Only Numbers*, Hyperion, New York, 1998.

R. Honsberger, *Ingenuity in Mathematics*, Mathematical Association of America, Washington, 1970.

———, *Mathematical Gems*, Mathematical Association of America, Washington, 1973.

———, *Mathematical Gems II*, Mathematical Association of America, Washington, 1976.

———, *Mathematical Morsels*, Mathematical Association of America, Washington, 1978.

———, *More Mathematical Morsels*, Mathematical Association of America, Washington, 1991.

———, *Episodes in Nineteenth and Twentieth Century Euclidean Geometry*, Mathematical Association of America, Washington, 1995.

T. Hull, On the mathematics of flat origamis, *Congressus Numerantum*, 100 (1994), pp. 215–224.

———, Origami quiz, *Mathematical Intelligencer*, 26 (2004), pp. 38–39, 61–63.

G. Hungerbühler, Proof without words: The triangle of medians has three-fourths the area of the original triangle, *Mathematics Magazine*, 72 (1999), p. 142.

H. Huzita, Understanding geometry through origami axioms. *Proceedings of the First International Conference on Origami in Education and Therapy* (COET91), J. Smith ed., British Origami Society (1992), pp. 37–70.

R. A. Johnson, A circle theorem. *American Mathematical Monthly*, 23 (1916), pp. 161–162.

R. F. Johnsonbaugh, Another proof of an estimate for *e*. *American Mathematical Monthly* 81 (1974), pp. 1011–1012.

W. Johnston and J. Kennedy, Heptasections of a triangle, *Mathematics Teacher*, 86 (1993), p. 192.

J. P. Jones and S. Toporowski, Irrational numbers. *American Mathematical Monthly* 80 (1973), pp. 423–424.

D. E Joyce, *Euclid's Elements*, http://aleph0.clarku.edu/~djoyce/java/elements/ elements.html, 1996.

D. Kalman, Six ways to sum a series, *College Mathematics Journal*, 24 (1993), pp. 402–421.

K. Kawasaki, Proof without words: Viviani's theorem, *Mathematics Magazine*, 78 (2005), p. 213.

N. D. Kazarinoff, *Geometric Inequalities*, Mathematical Association of America, Washington, 1961.

J. B. Keiper, Stirling's formula improved, *Two-Year College Mathematics Journal* 10 (1979). pp. 38–39.

L. M. Kelly and W. O. J. Moser, On the number of ordinary lines determined by *n* points. *Canadian Journal of Mathematics*, 10 (1958), pp. 210–219.

A. B. Kempe, *How to Draw a Straight Line: A Lecture on Linkages*, Macmillan and Company, London, 1877.

R. B. Kerschner, On paving the plane, *APL Technical Digest*, 8 (1969), pp. 4–10.

D. A. Klarner, *The Mathematical Gardner*, Prindle, Weber, and Schmidt, Boston, 1981.

M. S. Knebelman, An elementary limit, *American Mathematical Monthly*, 50 (1943), p. 507.

S. H. Kung, Sum of squares, *College Mathematics Journal*, 20 (1989), p. 205.

———, Proof without words: Every triangle has infinitely many inscribed equilateral triangles, *Mathematics Magazine* 75 (2002), p. 138.

J. Kürschak, Über das regelmässige Zwölfeck, *Math. naturw. Ber. Ung.* 15 (1898), pp. 196–197.

G. Lamé, Un polygone convexe étant donné, de combien de manières peut-on le partager en triangles au moyen de diagonals? *Journal de Mathématiques Pures et Appliquées*, 3 (1838), pp. 505–507.

R. J. Lang, *Origami and Geometric Constructions*, http://langorigami.com/ science/hha/origami_constructions.pdf, 2003.

L. Larson, A discrete look at $1 + 2 + \cdots + n$, *College Mathematics Journal*, 16 (1985), pp. 369–382.

W. G. Leavitt, The sum of the reciprocals of the primes, *Two-Year College Mathematics Journal*, 10 (1979), pp. 198–199.

M. Livio, *The Golden Ratio: The Story of Phi, the World's Most Astonishing Number*, Broadway Books, New York, 2002.

E. S. Loomis, *The Pythagorean Proposition*, National Council of Teachers of Mathematics, Reston, VA, 1968.

S. Loyd, *Sam Loyd's Cyclopedia of 5000 Puzzles, Tricks, and Conundrums (With Answers)*, The Lamb Publishing Co., New York, 1914. Available online at http://www.mathpuzzle.com/loyd/.

W. Lushbaugh, (no title), *Mathematical Gazette*, 49 (1965), p. 200.

D. MacHale, $\mathbb{Z} \times \mathbb{Z}$ is a countable set, *Mathematics Magazine*, 77 (2004), p. 55.

P. R. Mallinson, Proof without words: Area under a polygonal arch, *Mathematics Magazine*, 71 (1998a), p. 141.

———, Proof without words: The length of a polygonal arch, *Mathematics Magazine*, 71 (1998b), p. 377.

E. Maor, *e: The Story of a Number*, Princeton University Press, Princeton, 1994.

E. A. Margerum and M. M. McDonnell, Proof without words: Construction of two lunes with combined area equal to that of a given right triangle, *Mathematics Magazine*, 70 (1997), p. 380.

M. Moran Cabre, Mathematics without words, *College Mathematics Journal*, 34 (2003), p. 172.

L. J. Mordell and D. F. Barrow, Solution to Problem 3740, *American Mathematical Monthly*, 44 (1937), pp. 252–254.

C. Mortici, Folding a square to identify two adjacent sides, *Forum Geometricorum* 9 (2009), pp. 99–107.

P. J. Nahin, *An Imaginary Tale: The Story of i*, Princeton University Press, Princeton, 1998.

F. Nakhli, The vertex angles of a star sum to 180°, *College Mathematics Journal*, 17 (1986), p. 238.

NCTM, *Principles and Standards for School Mathematics,* National Council of Teachers of Mathematics, Reston, VA, 2000.

R. B. Nelsen, *Proofs Without Words: Exercises in Visual Thinking*, Mathematical Association of America, Washington, 1993.

————, *Proofs Without Words II: More Exercises in Visual Thinking*, Mathematical Association of America, Washington, 2000.

————, Proof without words: The area of a salinon, *Mathematics Magazine,* 75 (2002a), p. 130.

————, Proof without words: The area of an arbelos, *Mathematics Magazine,* 75 (2002b), p. 144.

————, Proof without words: Lunes and the regular hexagon, *Mathematics Magazine,* 75 (2002c), p. 316.

————, Mathematics without words: Another Pythagorean-like theorem, *College Mathematics Journal*, 35 (2004), p. 215.

I. Niven, A simple proof that π is irrational, *Bulletin of the American Mathematical Society* 53 (1947), p. 509.

————, Convex polygons that cannot tile the plane, *American Mathematical Monthly*, 85 (1978), pp. 785–792.

————, *Maxima and Minima Without Calculus*, Mathematical Association of America, Washington, 1981.

R. L. Ollerton, Proof without words: Fibonacci tiles, *Mathematics Magazine*, 81 (2008), p. 302.

J. O'Rourke, Folding polygons to convex polyhedra, in T. V. Craine and R. Rubenstein (eds.), *Understanding Geometry in a Changing World*, National Council of Teachers of Mathematics, Reston, VA, 2009.

H. Ouellette and G. Bennett, The discovery of a generalization: An example in problem solving, *Two-Year College Mathematics Journal*, 10 (1979), pp. 100–106.

R. Paré, A visual proof of Eddy and Fritsch's minimal area property, *College Mathematics Journal*, 26 (1995), pp. 43–44.

T. Peter, Maximizing the area of a quadrilateral, *College Mathematics Journal*, 34 (2003), pp. 315–316.

J. M. H. Peters, An approximate relation between π and the golden ratio, *Mathematical Gazette*, 62 (1978), pp. 197–198.

K. Pinter, Proof without words: The area of a right triangle, *Mathematics Magazine*, 71 (1998), p. 314.

G. Pólya, *Mathematical Discovery: On Understanding, Learning, and Teaching Problem Solving*, John Wiley & Sons, New York, 1965.

————, *Let Us Teach Guessing* (DVD), The Mathematical Association of America, Washington, 1966.

S. Portnoy, A Lewis Carroll Pillow Problem: Probability of an obtuse triangle, *Statistical Science*, 9 (1994), pp. 279–284.

M. M. Postnikov and A. Shenitzer, The problem of squarable lunes, *American Mathematical Monthly*, 107 (2000), pp. 645–651.

F. Pouryoussefi, Proof without words, *Mathematics Magazine*, 62 (1989), p. 323.

A. D. Rawlins, A note on the golden ratio, *Mathematical Gazette*, 79 (1995), p. 104.

P. Ribenboim, *The Little Book of Bigger Primes*, 2nd ed., Springer, New York, 2004.

J. F. Rigby, Equilateral triangles and the golden ratio, *Mathematical Gazette*, 72 (1988), pp. 27–30.

D. Rubinstein, Median proof, *Mathematics Teacher*, 96 (2003), p. 401.

Y. Sagher, Counting the rationals, *American Mathematical Monthly*, 96 (1989), p. 823.

F. Saidak, A new proof of Euclid's theorem, *American Mathematical Monthly*, 113 (2006), pp. 937–938.

D. Schattschneider, *M. C. Escher: Visions of Symmetry*, Harry N. Abrams, Publishers, New York, 2004.

———, Beauty and truth in mathematics, in: *Mathematics and the Aesthetic: New Approaches to an Ancient Affinity*, N. Sinclair, D. Pimm, W. Higgenson (editors), Springer, New York, 2006, pp. 41–57.

N. Schaumberger, Alternate approaches to two familiar results. *College Mathematics Journal*, 15 (1984), pp. 422–423.

R. Schmalz, *Out of the Mouths of Mathematicians*, Mathematical Association of America, Washington, 1993.

W. H. Schultz, An observation, *American Mathematical Monthly*, 110 (2003), p. 423.

B. Schweizer, Cantor, Schröder, and Bernstein in orbit, *Mathematics Magazine*, 73 (2000), pp. 311–312.

D. Singmaster, Covering deleted chessboards with dominoes, *Mathematics Magazine*, 48 (1975), pp. 59–66.

S. L. Snover, Four triangles with equal area. In R. B. Nelsen, *Proofs Without Words II*, Mathematical Association of America, Washington, 2000; p. 15.

J. M. Steele, *The Cauchy-Schwarz Master Class*, Mathematical Association of America and Cambridge University Press, Washington-Cambridge, 2004.

S. K. Stein, Existence out of chaos, in R. Honsberger (ed.), *Mathematical Plums*, Mathematical Association of America, Washington, 1979, pp. 62–93.

H. Steinhaus, *Mathematical Snapshots*, Oxford University Press, New York, 1969.

E. Steinitz, Polyeder und Raumeinteilungen, *Enzykl. Math. Wiss.* 3 (1922) Geometrie, Part 3AB12, pp. 1–139.

P. Strzelecki and A. Schenitzer, Continuous versions of the (Dirichlet) drawer principle, *College Mathematics Journal*, 30 (1999), pp. 195–196.

A. Sutcliffe, A note on the sum of squares, *Mathematics Magazine*, 36 (1963), pp. 221–223.

A. E. Taylor and W. R. Mann, *Advanced Calculus*, 2nd ed., Xerox College Publishing, Lexington MA, 1972.

M. G. Teigen and D. W. Hadwin, On generating Pythagorean triples, *American Mathematical Monthly*, 78 (1971), pp. 378–379.

C. W. Trigg, A hexagonal configuration, *Mathematics Magazine*, 35 (1962), p. 7.

———, Solution to Problem 852, *Mathematics Magazine*, 46 (1973), p. 288.

J. van de Craats, The golden ratio from an equilateral triangle and its circumcircle, *American Mathematical Monthly*, 93 (1986), p. 572.

C. Vanden Eynden, Proofs that $\sum 1/p$ diverges, *American Mathematical Monthly*, v. 87 (1980), pp. 394–397.

D. E. Varberg, Pick's theorem revisited, *American Mathematical Monthly*, 92 (1985), pp. 584–587.

T. Walker, A geometric telescope, *American Mathematical Monthly*, 109 (2002), p. 524.

H. Walser, *The Golden Section*, Mathematical Association of America, Washington, 2001.

A. J. B. Ward, Divergence of the harmonic series, *Mathematical Gazette*, v. 54 (1970), p. 277.

D. Wells, *The Penguin Dictionary of Curious and Interesting Geometry*, Penguin Books, London, 1991.

A. M. Yaglom and I. M. Yaglom, *Challenging Mathematical Problems with Elementary Solutions*, vol. 1, Dover Publications, Inc., New York, 1964.

Index

About the Authors

Claudi Alsina was born on 30 January 1952 in Barcelona, Spain. He received his B.A. and Ph.D. in mathematics from the University of Barcelona. His post-doctoral studies were at the University of Massachusetts, Amherst. As Professor of Mathematics at the Technical University of Catalonia, he has developed a wide range of international activities, research papers, publications and hundreds of lectures on mathematics and mathematics education. His latest books include *Associative Functions: Triangular Norms and Copulas* with M.J. Frank and B. Schweizer, WSP, 2006; *Math Made Visual. Creating Images for Understanding Mathematics* (with Roger B. Nelsen), MAA, 2006; *When Less is More: Visualizing Basic Inequalities* (with Roger B. Nelsen), MAA, 2009; *Vitaminas Matemáticas* and *El Club de la Hipotenusa*, Ariel, 2008, *Geometria para Turistas*, Ariel, 2009.

Roger B. Nelsen was born on 20 December 1942 in Chicago, Illinois. He received his B.A. in mathematics from DePauw University in 1964, and his Ph.D. in mathematics from Duke University in 1969. Roger was elected to Phi Beta Kappa and Sigma Xi and taught mathematics and statistics at Lewis & Clark College for forty years before his retirement in 2009. His previous books include *Proofs Without Words: Exercises in Visual Thinking,* MAA, 1993; *An Introduction to Copulas,* Springer 1999 (2nd. ed. 2006); *Proofs Without Words II: More Exercises in Visual Thinking*, MAA, 2000; *Math Made Visual: Creating Images for Understanding Mathematics* (with Claudi Alsina), MAA, 2006; *When Less is More: Visualizing Basic Inequalities* (with Claudi Alsina) MAA, 2009; and *The Calculus Collection: A Resource for AP and Beyond* (with Caren Diefenderfer, editors), MAA, 2010.